Emergency and Security Lighting

Emergency and Security Lighting

Gerard Honey

Newnes

OXFORD AUCKLAND BOSTON JOHANNESBURG MELBOURNE NEW DELHI

VCTC Library
Hartness Library
Vermont Technical College
Randolph Center, VT 05061

Newnes
An imprint of Butterworth-Heinemann
Linacre House, Jordan Hill, Oxford OX2 8DP
225 Wildwood Avenue, Woburn, MA 01801-2041
A division of Reed Educational and Professional Publishing Ltd

A member of the Reed Elsevier plc group

First published 2001

© Gerard Honey 2001

All rights reserved. No part of this publication may be reproduced in any material form (including photocopying or storing in any medium by electronic means and whether or not transiently or incidentally to some other use of this publication) without the written permission of the copyright holder except in accordance with the provisions of the Copyright, Designs and Patents Act 1988 or under the terms of a licence issued by the Copyright Licensing Agency Ltd, 90 Tottenham Court Road, London, England W1P 0LP. Applications for the copyright holder's written permission to reproduce any part of this publication should be addressed to the publishers

British Library Cataloguing in Publication Data
A catalogue record for this book is available from the British Library

ISBN 0 7506 5037 0

Typeset in India at Integra Software Services Pvt Ltd, Pondicherry 605 005
Printed and bound in Great Britain by Biddles Ltd, Guildford and King's Lynn

Contents

Preface

New workplace directives and stringent building regulations require more premises to install emergency lighting systems. Amendments to the fire precautions (workplace) regulations invoke additional needs for written fire risk assessment and emergency plans for premises including emergency lighting to form a part of the means of escape. Local authority bylaws also oblige the updating of many premises holding certificates under earlier Housing Acts. As the safety of the occupants of buildings is becoming a major priority with the introduction of new harmonized standards the need for emergency lighting is increasingly apparent. There is therefore a cause to install emergency lighting in most factories, buildings open to the public and those for multiple occupancy. In addition there is provision under the Fire Precautions Act to extend its use into hotels and boarding houses without further legislation.

Security lighting is being covered in the same title as emergency lighting because it also has a safety aspect. This is in addition to its use as a deterrent to crime. Lighting of this nature may be automatically selected on demand or be used over extended periods of time to provide illumination at all times when natural light is unavailable. It may equally be used to supplement equipment employed for surveillance purposes.

Gerard Honey

Acknowledgements

The author would like to thank the following for information received:

Cooper Lighting and Security Ltd
Emergi-Lite Safety Systems Ltd
Existalite Ltd
TAVCOM Training Ltd

Part 1 Emergency Lighting Systems

1 Introduction

As the safety of occupants in buildings is becoming a major priority with the introduction of new harmonized standards the need for addressing emergency lighting alongside the fire risk assessment is becoming more apparent. We are obliged to be alert to a part of the written fire risk assessment and emergency plan of the premises for any business covering the means of escape in the building and the identification of hazards. This is governed by law which requires us to:

- Carry out a fire risk assessment of the workplace to consider all employees, the public, disabled people and people with special needs.
- Identify the significant findings and people at risk. These must be recorded if more than five people are employed.
- Provide and maintain fire precautions.
- Provide information, instruction and training to employees.
- Nominate people to undertake special roles under the emergency plan and have an undertaking that employees will co-operate to ensure that the workplace is safe from fire and its effects.

Any business employing more than five persons is required by the amendments to the Fire Precautions (Workplace) Regulations 1997 to carry out this fire risk assessment and identify the significant findings and people at risk. From this there must be a programme to correct any failings or shortcomings. Inherent in this assessment is a means of escape from the building and the illumination of escape routes.

The Health and Safety at Work Act places a duty on all employers and occupiers of buildings to provide a safe environment which has of itself led fire officers to insist on the installation of emergency lighting in most factories, buildings open to the public and those for multiple occupancy. The Fire Precautions Act has historically been regarded as an embodying Act so there was provision for it to be extended beyond its particular course for hotels and boarding houses without further legislation. For this reason, as a building's fire protection, other measures are assessed alongside this leading to a need to update, upgrade or indeed install emergency lighting. The inference from the Health and Safety at Work Act 1974 is that if an accident were to occur that could have been prevented if adequate emergency lighting had been available the persons responsible for the building safety could be liable for prosecution.

The Fire Precautions in the Workplace Regulations 1997 used and modified the Fire Precautions Act of 1971 stipulating that any site with five or more persons must keep a formal record of the Fire Risk Assessment in evaluating the site to ensure its safety. Compliance with BS 5839 for fire detection and BS 5266 for emergency lighting are used as the basis for meeting the requirements of this. There are other legal requirements and the Building Regulations 1991 list the areas in which emergency lighting is required and the need to meet BS 5266/ LP 1007 for standards of installation. In addition some premises operating as places of entertainment, including those that sell alcohol, need to be licensed or given legislative acceptance by the local authority. In general this is controlled by a satisfactory report on the site by the local fire authority.

It will be found that some specific sites are excluded from the Workplace Regulations because they present specific risks and in this we include such areas as mines and construction sites. These, however, do have their own regulations and in the case of shortcomings in some areas of fire protection these can be compensated for by improved levels of emergency lighting and the actual fire alarm systems.

It is also clear that the logic of providing emergency lighting extends well beyond that of aiding emergency evacuation in the event of a fire. This is because it provides safety for something as simple and frequent as a power cut or the disruption of the normal lighting of a particular circuit or area of a premises. It effectively comes on to illuminate an area and remains energized when the electrical supply to the normal system fails. This is done by means of battery cells. It is therefore seen as a safety system for evacuation purposes but is not to be confused with standby lighting that is a form of emergency lighting provided to enable activities to continue in the normal sense under mains failure conditions. In practice standby lighting makes use of standard luminaires although there is an issue of identifying particular luminaires that are to operate under standby conditions by the selecting of low power consumption control gear circuits in order to maximize the light provided against the power consumed.

We come to accept that there is an ongoing need for emergency lighting and without it there is a risk of death, injury and panic particularly in areas where persons are unfamiliar with their surroundings and the building layout. This is more critical where hazardous practices are being carried out involving such activities as moving machinery.

There are laws saying that we must implement emergency lighting plus guides and codes of practice which both interpret and implement legislation. There are areas which in general are exempt from legislation or the need for a fire certificate but in the main this only applies to premises such as private dwellings and small shops operated by the self-employed. For the areas in which legislation applies, although fire certificates are not required, the owners must carry out a fire precautions risk assessment and if five or more people are employed there is a legal requirement to record the significant findings of the risk and document the measures taken to cope with the risks. For those areas requiring fire certificates this will be carried out by the inspecting body such as the local fire authority. For those areas not needing fire certificates or falling under the Fire Precautions (Workplace) Regulations 1997 these will be governed by legislation for their specific role.

The Fire Precautions (Workplace) Regulations state that 'Emergency routes and exits must be indicated by signs and emergency routes and exits requiring illumination shall be provided with emergency lighting of adequate intensity in case of failure of the normal lighting.' The Building Regulations require every emergency exit route to be adequately lit with emergency lighting in the event of the failure of its normal source. Therefore the essential design objective of emergency lighting when referring to escape lighting is to indicate the escape routes, provide illumination along such routes and towards the exit routes, ensure that fire alarm call points and fire-fighting equipment can be located and to permit operations concerned with safety features. There are also needs for exit signs to be located so that the specific exits and the directions to it are easily recognized and these are to be at any location where the route may be in doubt.

In practice emergency lighting luminaires can be matched to the application so that elegant low profile units are used to blend in with the existing décor or other versions can be recessed, decorative, vandal resistant or weatherproof. The term luminaire is used throughout the industry to describe the apparatus which distributes, filters and transforms the lighting given by a lamp or lamps. It includes all of the components that are necessary to fix and to protect the lamps and the parts that are needed to connect these into the supply circuit. It will also be found that internally illuminated signs are regarded as a special form of luminaire.

The control of luminaires and the cabling can be configured so that they may be switched on and off for normal duty when the mains is healthy but if the mains fail they can then operate from the standby batteries.

The choice of system is always very much dictated by the use of the building and its size and architecture.

In this book we first look at conventional emergency lighting networks and then later direct our attention to low mounted way guidance systems that use a special type of illumination in the form of strip lighting.

For conventional systems there are two essential types, namely self-contained or central battery. The former category uses single point luminaires that contain all of the elements including the battery and lamp in one unit. These are not complex and have low initial installation costs. Central battery systems derive their power from a central point and are used in large installations as maintenance and operating costs in the long term are lower than those of the self-contained system.

The importance of correctly specified emergency lighting systems cannot be underestimated. An awareness of its vital safety and security function is constantly being raised by developers and architects and more stringent regulations are progressively coming into force. The technology of the subject is also making rapid advances leading to a range of products that have refined system techniques to such a degree that investment costs now make it an even more attractive proposition for integration into other building facilities.

The need to consider emergency lighting as a priority subject is necessary if we are not to encounter fatal consequences should smoke or the failure of normal lights prevent building occupants from being able to find the escape routes. The way in which the emergency lighting is installed, specified and maintained is very much governed by many factors and these are all considered in the content of this book. There are also selection techniques that relate to the wiring since different requirements exist for the various systems and there is a need to understand the characteristics of the batteries as these form an essential part of any emergency lighting programme.

The luminaires that are used must also blend into the environment, so in addition to having an adequate level of photometric performance they must match the environment, for in many instances aesthetics are important. Therefore we can conclude that the correct decision with respect to the most efficient and cost-effective system that is to be installed is needed at the outset. If this is not achieved there will surely be unnecessary expenditure at a later date in the building programme and the maintenance of the premises during its working life.

2 System selection

The selection of the system is very much governed by legislation and the standards that are currently in force. The main legislation for emergency lighting is:

- Fire Precautions (Workplace) Regulations 1997
- The Fire Precautions Act 1971
- Health and Safety at Work Act 1974
- Cinematographic Act 1952
- Cinematographic (Safety) Regulations 1955 No. 1129
- The Workplace Directive

There are also various bylaws and other issues of legislation that apply and some larger local authorities produce their own standards based on time honoured legislation.

In addition there are a number of European Standards being produced that will become part of future UK legislation. It is important to note also the role of the Industry Committee for Emergency Lighting (ICEL) which has a certification and registration scheme for both luminaires and conversion modules. By using their registered products all installers can be assured that they are purchasing quality components.

Industry standards also apply and ICEL 1001 1978 provides a nationwide product standard to formulate BS 4533.102.22 and IEC 598.2.22.

The most common standards used in the actual design are BS 5266 Parts 1 and 7: 1999. These provide a method for designing an emergency lighting scheme in a building and are recognized as the essential design codes although certain governing bodies may include deviations from these codes and include additional requirements. There is also a need to be alert to the importance of low mounted way guidance systems and how they differ from the more widely known emergency lighting practices. This technique is considered in its own right in Chapter 8 alongside ancillary applications.

The practical approach to the selection of the traditional type of system is to determine the purpose of the lighting and whether it is to operate as escape lighting or standby lighting.

In the first instance we can define the actual emergency lighting as lighting provided for use when the supply to the normal lighting fails. This disruption can be caused by a power cut, rupture of a fuse, disconnection of a circuit by a protective device or failure of the cabling to the normal lighting. It may also be deliberately induced by persons attempting to cause confusion in a building. Emergency lighting hence helps prevent injury to the occupants of a building plus threats to property.

The escape lighting is that part of the emergency lighting which is provided to ensure that the escape route is illuminated at all material times. The escape route being the route forming part of the means of escape from a point in a building to a final exit.

Standby lighting must not be confused with escape lighting because it is provided to enable normal activities to continue when the mains supply fails. We can therefore say that emergency lighting is installed to operate from battery cells when the mains supply fails and is intended primarily for safety, whereas standby lighting is used to enable the normal occupation of a building. It should be noted that standby lighting may be needed in areas in which activities must continue so it is not always part of the emergency lighting scheme.

If the standby lighting is used as emergency lighting it must meet its requirements or a full emergency lighting network should be added to the premises alongside the standby lighting.

In accordance with the European Standards there are four systems in use with emergency lighting:

- Self-contained systems.
- Small central systems.
- Large central battery systems.
- Central inverter systems.

2.1 Self-contained systems

These systems use single point or self-contained luminaires. They contain all of the elements within one unit to include the lamp, battery, charging electronics and mains failure indicator usually in the form of an LED. They may also have an inverter to operate a fluorescent tube.

2.2 Small central systems

These are compact systems using sealed lead–acid batteries supplying a local area with an emergency supply. There may be a number of these systems in a building.

2.3 Large central battery systems

These consist of a large battery network and can occupy a particular purpose-built room or storage area.

2.4 Central inverter systems

The central inverter is used to provide mains power to the emergency circuit by converting the central battery voltage to ac. These systems use standard mains luminaires so there is a great choice of lamps although consideration must be given to mains lamps operating at full power. Luminaires that operate at high efficiency with high light output must be used.

The system selection must therefore account for the legislation in force for the premises and the choice between the four system types.

At first we can consider the selection in general terms and then go on to look at the four different types in more detail at a later stage.

Self-contained systems are inexpensive, easy to install and readily adapted. They are cabled in PVC insulated and sheathed cable and confine any failure to a single luminaire within which the integral battery is held. These systems tend to be used in small buildings.

Small central systems and large central battery systems provide their power from a given point either to emergency luminaires in a single lighting circuit or to normal light fittings in an entire building. These units do not have integral batteries but the standby batteries are held at a remote location. The luminaires are referred to as 'slaves'. Large central battery systems providing dc power tend to be used in buildings in which there is a need for a large number of luminaires and the premises can have a varied and long life span. Maintenance and operating costs in these situations is lower in the long term than that experienced with self-contained systems. However, the installation of dc central battery systems is initially more expensive and if there is a need to use normal lighting units they must be fitted with conversion kits for acceptance of dc power. The system selection is therefore between higher initial investment costs and lower maintenance charges, or a lesser investment cost but with a budget to cover the replacement of batteries and the higher cost of maintenance over the life of the building.

The option is to use ac systems with inverters which do not require normal mains lighting units to be converted. In addition unlike converted units the luminaire provides full light output. When an ac emergency lighting unit is installed the lamps have a combined emergency lighting supply from the inverter and the normal mains supply. The luminaire can be configured in such a way that it can be switched on and off during normal operation and to achieve this a changeover device is used. During normal mains conditions the local lighting circuit provides the power whereas the inverter supplies the power in the event of a mains failure. The advantage of both dc and ac central battery systems is that it enables buildings to be used to their maximum potential without a need to move luminaires to satisfy changing applications.

Central battery systems have good future prospects as battery life can be in the order of 25 years, maintenance is low and they are suitable for integration into other building facilities.

We can say that the choice between self-contained, small central and large central battery and inverter systems will always be governed by the assessment of equipment against the installation and maintenance costs. In general self-contained systems are most suited to small installations although the batteries have a short life and need to be changed after four years. In the case of a large number of luminaires being needed the equipment cost of a central battery system can be lower than that of the self-contained type but minimal maintenance is required. The small central system uses sealed lead–acid batteries which have a typical life of four to seven years whilst the large central battery system has a long design life from ten to 25 years.

The advantage of central inverter systems is that standard mains luminaires are used so the choice is not restricted and high light outputs are provided for greater spacing distances.

The four system types are covered in detail together with their advantages and disadvantages as we encounter them throughout the course of the book.

3 Design considerations

BS 5266, which relates to the emergency lighting of premises other than cinemas and certain other premises used for entertainment, refers to the provision of escape lighting and the design objective of fulfilling certain functions. Escape lighting can be defined as that part of the emergency lighting which is provided to ensure that the escape route is illuminated at all material times. The escape route is that route forming part of the means of escape from a point in a building to a final exit.

The requirement of the design is that when the supply to the normal lighting or part of the normal lighting in occupied premises fails escape lighting must fulfil the following functions:

- Indicate clearly and unambiguously the escape routes. This calls for the use of exit signs. These should comply with the format of BS 5499 in addition to the Running Man pictogram format as specified by the European Signs Directive. Exit signs should be visible from any point on an escape route.
- Provide illumination along escape routes to allow safe movement towards and through the exits by luminaires spaced at appropriate distances.
- Ensure that fire alarm call points and fire-fighting equipment provided along escape routes can be readily located.
- Permit operations concerned with safety measures.

The design can be formulated by systematically using a policy of design stage considerations namely:

- Points of emphasis.
- Defined escape routes – level of illumination.
- Essential areas – luminaire locations.
- Photometric design.
- Anti-panic/open core areas – level of illumination.
- High risk task area lighting.
- Exit signs – format and size.

When in the process of considering these design stages it is to be noted that luminaires are to be mounted at a distance of at least 2 m above the floor level. There is no upper limit quoted but if there is a significant risk of smoke affecting the illuminance across the floor the luminaires are then to be applied below the expected smoke level. If the risk of smoke is high then low level way guidance systems should be installed. These are covered in Chapter 8. It is also to be noted in all cases that it is recommended that a larger number of low powered luminaires are used in preference to a lesser number of more powerful devices to cater for luminaire failure.

3.1 Points of emphasis

An essential part of the design procedure is to provide a method for designing an emergency lighting scheme by installing luminaires at specific points as Figure 3.1.

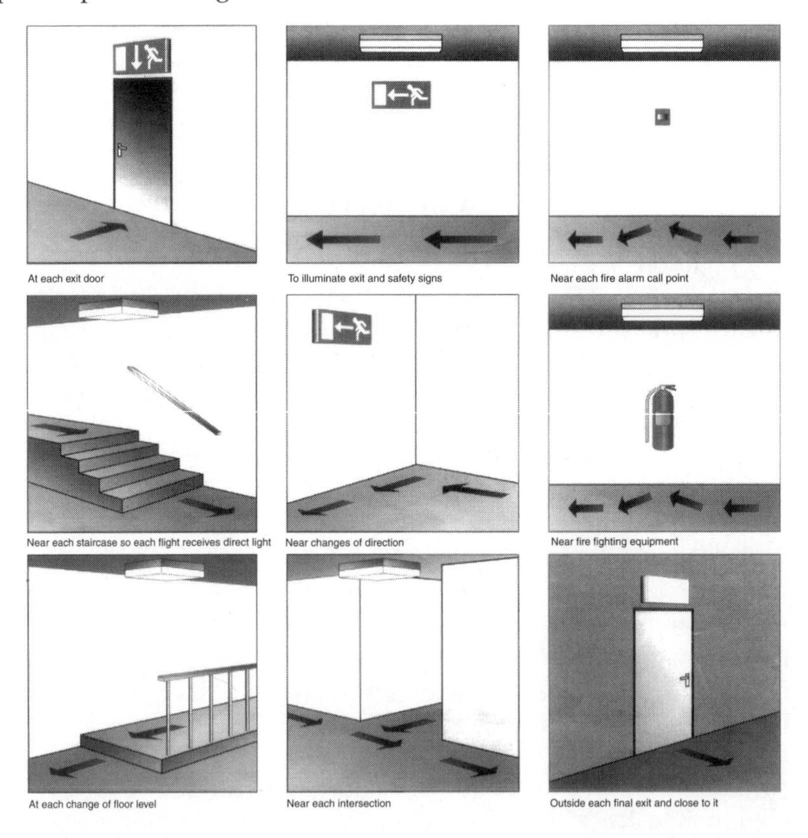

Figure 3.1 *Points of emphasis*

The first part of the design procedure is to determine the siting and positions of luminaires to account for specific hazards and to highlight all safety equipment and directional signs. This is to be performed regardless of whether the luminaires are sited on an emergency escape route or in an anti-panic area. An anti-panic area of lighting is part of the emergency escape lighting which aims to reduce panic and provide illumination and direction finding to enable people to reach escape routes in a safe manner.

Certain points are mandatory and referred to as 'points of emphasis'.

These points of emphasis should be illuminated by a luminaire and exit directional sign:

- At each exit door that is intended to be used in an emergency.
- To illuminate exit and safety signs required by the enforcing authority.
- Near each staircase so that every flight of stairs receives direct light.
- Near any other change of floor level.
- Near each change of direction (other than on a staircase).
- Near each intersection of corridors.
- Near every fire alarm call point.
- Near every unit of fire-fighting equipment.
- Outside each final exit and close to it.

As an addition to the foregoing luminaires should also be placed near to any first aid points.

For the purposes of 'near' in relation to first aid points and other points of emphasis this is normally considered to be within 2 m measured in a horizontal direction.

We can overview escape routes and exits as those being the most direct, unambiguous and practical ways to the final exit points.

The emergency lights to illuminate the route must take into account any hazard points, fire-fighting and safety equipment and be placed within 2 m of any particular discontinuity on the route. The exit path in practice is regarded as a swath 2 m wide with wide routes considered as a multiple of this. In arranging the luminaires there must be a uniform illumination of no greater than 40:1. This is best achieved with fittings that support an even light distribution over a wide aspect. In the event that the escape route crosses an open area the exact circumstances must be assessed because storage spaces and pathways may require an amendment in escape routing.

In all cases the users of the emergency lighting are to be able to distinguish colours to include red fire alarm equipment and safety signs in green. It should be understood that low pressure sodium

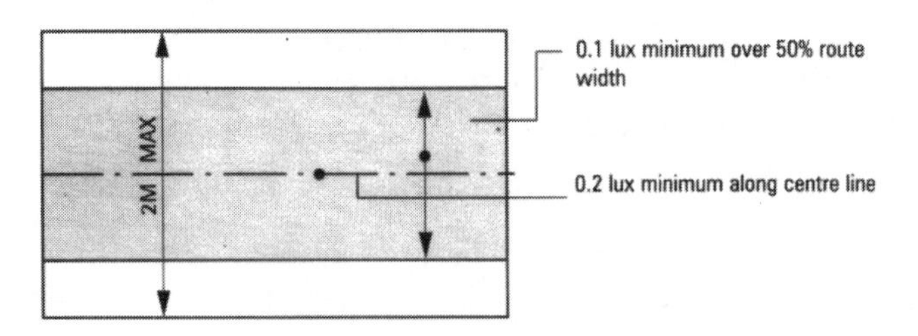

Figure 3.2 *Clearly defined escape routes*

lamps in particular have poor colour rendering properties but most products manufactured for the mainstream emergency lighting market do satisfy the lowest datum level of Ra40.

3.2 Defined escape routes – level of illumination

Once the points of emphasis have been covered there is a need to consider the provision of any additional luminaires to ensure that minimum levels of illuminance are met so that all escape routes can be used safely. This is to account for minor hazards such as steps and objects that may be in troublesome positions. Every compartment on the escape route must have at least two luminaires in the event that a luminaire may fail.

It is recommended that there is a minimum 1 lux along the centre line of the route and 0.5 lux over a 1 m wide central band if route obstructions are present or the route is liable to be used by old people. This can be relaxed to 0.2 lux along the centre line and 0.1 lux over a 1 m wide central band for unobstructed routes of low risk according to Figure 3.2.

For escape routes up to 2 m wide the centre line illuminance at floor level must not fall below 0.1 lux. Wider routes are treated as a number of 2 m wide bands with the values applied to each band. Uniformity should not be less than a 40:1 ratio.

3.3 Essential areas – luminaire locations

Although not part of the escape route certain other additional areas require the use of emergency lighting because they are classed as risk areas. Luminaires should therefore be installed at the locations as shown at Figure 3.3.

(a) Lift cars, although only in exceptional circumstances will they be part of the escape route, they do present a problem that the public may be trapped in them in the event of a supply failure.

(b) Toilets with facilities exceeding 8 m^2 floor area and all toilets for the disabled.

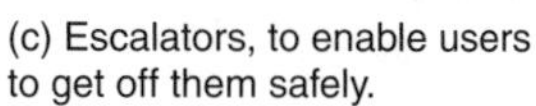

(c) Escalators, to enable users to get off them safely.

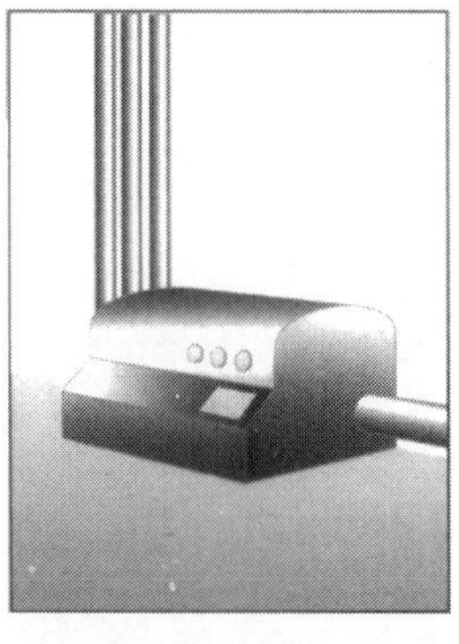

(d) Motor generator, control or plant rooms require battery supplied emergency lighting to assist any maintenance of operating personnel in the event of failure.

(e) Covered car parks: the normal pedestrian routes should be provided with non-maintained luminaires of at least 1 hour duration.

Figure 3.3 *Essential areas – luminaire locations*

These essential area locations are:

- Lifts. These are only used as part of the escape route in exceptional circumstances. Occupants, however, could be trapped within them in the event of a power failure.
- Toilets exceeding 8 metres square floor area.
- All toilets for the disabled.
- Escalators and moving walkways. This is to allow users to leave them safely.
- Generator, control and plant rooms. This is to assist operating or maintenance personnel in the event of failure.
- Pedestrian routes in covered car parks. Non-maintained luminaires of at least 1 hour duration are to be installed.

It is also practice to include toilets smaller than 8 metres square floor area unless they are illuminated by emergency lighting borrowed from another area plus small lobbies if they have no borrowed light.

3.4 Photometric design

In order to determine the required number of additional luminaires along an escape route, illumination spacing tables are provided by manufacturers on their product pages. These are needed to ensure that the correct levels of lighting and uniformity are achieved. We may also note that fewer fittings always save on installation costs.

It is these spacing tables that provide the information to establish if fittings are needed in addition to those for the points of emphasis. Figures 3.4(a) and 3.4(b) shows examples of spacing tables used to select appropriate luminaires and to provide 1 and 0.2 lux minimum illuminance levels.

3.5 Anti-panic/open core areas – level of illumination

Any area larger than 60 metres square or open areas with an escape route passing through it require emergency lighting. The levels of illumination required can differ between the governing standards but will either invoke:

1 lux average over the floor area

or 0.5 lux minimum throughout the floor area excluding a perimeter border of 0.5 m.

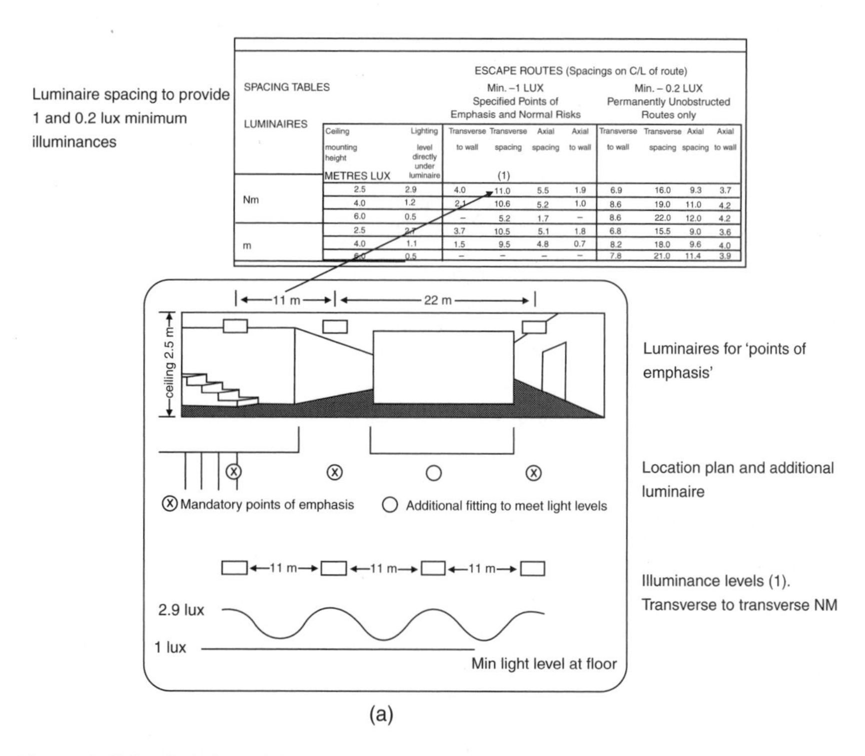

SPACING TABLES — ESCAPE ROUTES (Spacings on C/L of route)

LUMINAIRES	Ceiling mounting height METRES	Lighting level directly under luminaire LUX	Min. –1 LUX Specified Points of Emphasis and Normal Risks: Transverse to wall	Transverse spacing (1)	Axial spacing	Axial to wall	Min. – 0.2 LUX Permanently Unobstructed Routes only: Transverse to wall	Transverse spacing	Axial spacing	Axial to wall
Nm	2.5	2.9	4.0	11.0	5.5	1.9	6.9	16.0	9.3	3.7
	4.0	1.2	2.1	10.6	5.2	1.0	8.6	19.0	11.0	4.2
	6.0	0.5	–	5.2	1.7	–	8.6	22.0	12.0	4.2
m	2.5	2.7	3.7	10.5	5.1	1.8	6.8	15.5	9.0	3.6
	4.0	1.1	1.5	9.5	4.8	0.7	8.2	18.0	9.6	4.0
	6.0	0.5	–	–	–	–	7.8	21.0	11.4	3.9

Figure 3.4(a) *Spacing tables to show provision of 1 and 0.2 lux minimum illuminances*

The spacing tables that are provided by the manufacturers give simple and accurate data that can easily be used. The evidence of compliance with light levels is to be supplied by the system designer. The manufacturers' fittings are photometrically tested at an authorized test house and the spacing data is registered by the ICEL. Copies of the spacing data provide the necessary verification.

Certification is a method of approving the manufacture of emergency lighting equipment as monitored in the UK by the British Standards Institution (BSI). Any company that has the BSI Kitemark certification for their luminaires will have submitted each luminaire type individually to the BSI for approval and the manufacturing plant is then subject to ongoing follow-up checks to ensure continued compliance with the requirements of the certification scheme. ICEL certification is in addition to the BSI Kitemark and proves that luminaires for use in defined escape routes are fire retardant and that the photometric data has been verified by a third party.

In addition a satisfactory test of operation is required using time honoured system log-books with appropriate test forms. It must

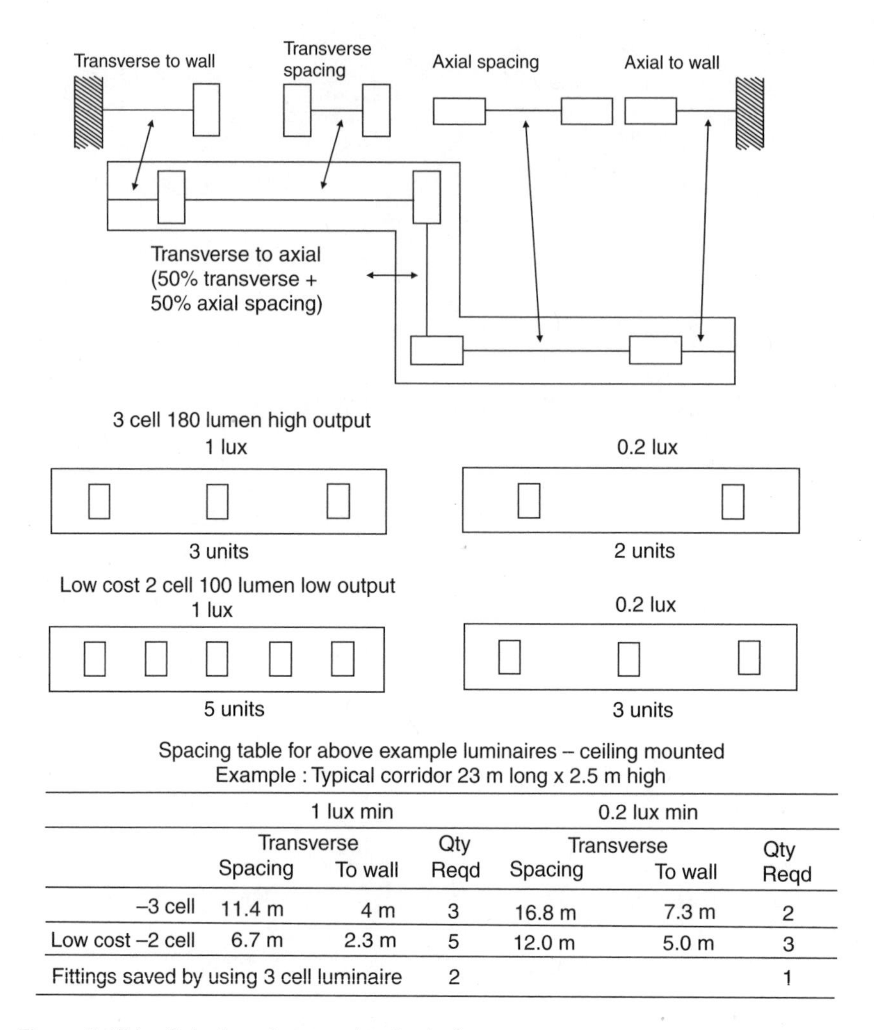

Spacing table for above example luminaires – ceiling mounted
Example : Typical corridor 23 m long x 2.5 m high

	1 lux min			0.2 lux min		
	Transverse Spacing	Transverse To wall	Qty Reqd	Transverse Spacing	Transverse To wall	Qty Reqd
–3 cell	11.4 m	4 m	3	16.8 m	7.3 m	2
Low cost –2 cell	6.7 m	2.3 m	5	12.0 m	5.0 m	3
Fittings saved by using 3 cell luminaire			2			1

Figure 3.4(b) *Selection of appropriate luminaires*

be understood that in order to ensure that the system remains at the correct level of illumination there is a need to formulate a system of maintenance. In general this is part of the normal test routine.

It is practice to design for systems that do not use a few large output luminaires but use a greater number of fittings. An example of photometric data tables for specific self-contained luminaires is shown in Figure 3.5. This illustrates two maintained and two non-maintained fittings identified as A–D.

It will be noted that the data table will allow for certain correction factors for the values quoted:

SELF-CONTAINED LUMINAIRES			2 m wide escape routes 1 lux min. along centre line 0.5 lux min. in 1 m central band				(Anti-panic) open areas 0.5 lux min. Luminaires arranged in a regular array				2 m wide clear and unobstructed escape routes 0.2 lux min. along centre line 0.1 lux min. in 1 m central band			
		Ceiling mounting height (m)	Transverse to wall	Transverse spacing	Axial to wall	Axial spacing	Transverse to wall	Transverse spacing	Axial to wall	Axial spacing	Transverse to wall	Transverse spacing	Axial to wall	Axial spacing
A	Maintained	2.5	1.8	5.6	1.5	4.7	2.1	5.6	1.7	4.6	4.2	10.9	3.6	8.9
		3	1.5	5.5	1.2	4.6	2.0	5.8	1.7	4.8	4.3	11.4	3.7	9.5
		4	–	–	–	–	1.7	5.8	1.5	4.9	4.5	12.1	3.9	10.4
		5	–	–	–	–	0.8	5.4	0.6	4.6	4.3	12.5	3.8	10.8
B	Non-maintained	2.5	2.0	6.0	1.7	5.1	2.2	6.0	1.9	4.8	4.2	10.9	3.6	8.9
		3	1.8	5.9	1.5	5.1	2.2	6.2	1.7	5.0	4.3	11.4	3.7	9.5
		4	–	–	–	–	2.0	6.3	1.7	5.3	4.5	12.1	3.9	10.4
		5	–	–	–	–	1.5	6.1	1.2	5.1	4.3	12.5	3.8	10.8
C	Maintained	2.5	2.0	5.9	1.7	5.1	2.1	5.6	1.8	4.7	4.4	11.2	3.7	9.2
		3	1.7	5.9	1.5	5.1	2.1	5.8	1.8	5.0	4.5	11.8	3.9	9.8
		4	–	–	–	–	1.8	5.8	1.6	5.1	4.7	12.7	4.0	10.7
		5	–	–	–	–	1.1	5.5	0.9	4.8	4.6	13.1	4.0	11.3
		6	–	–	–	–	–	–	–	–	4.3	13.2	3.7	11.5
D	Non-maintained	2.5	2.1	6.5	1.8	5.5	2.2	6.0	1.9	5.0	4.4	11.2	3.7	9.2
		3	2.1	6.6	1.8	5.7	2.2	6.2	1.9	5.3	4.5	11.8	3.9	9.8
		4	1.0	6.2	0.7	5.3	2.1	6.3	1.8	5.5	4.7	12.7	4.0	10.7
		5	–	–	–	–	1.6	6.2	1.4	5.4	4.6	13.1	4.0	11.3
		6	–	–	–	–	–	–	–	–	4.3	13.2	3.7	11.5

Figure 3.5 *Photometric data tables*

- Service (or maintenance) factor – This is a calculation that allows for the effect of dirt and ageing.
- K factor – This is the ratio between the light output from the lamp in the worst condition (normally at end of discharge and with any cable volt drop) to the light output at nominal voltage.
- Life (or maintained) factor – This is quoted for maintained lamps to allow for reduced output through the working life of the lamp.
- Volt drop factor – This will be quoted to cater for the design voltage declared by the manufacturer to which all of the ballast characteristics are related. The value is 80% to 90% of the maximum value of the rated voltage range.

These will be on the order of:

- Service (or maintenance factor) – 0.8
- K factor – 0.71 to 0.82 depending on luminaire type
- Life (or maintained factor) – 0.85 for maintained luminaires
- Volt drop factor – 0.90

Calculations for normal lighting can also be made by use of the formula

$$\text{Number of luminaires} = \frac{\text{lumen level required} \times \text{room area (length} \times \text{width)}}{\text{LDL (lamp output)} \times \text{UF (utilization factor)} \times \text{MF (maintenance factor)}}$$

The utilization factor is quoted at zero reflectance (UF0) and determines the proportion of light output from a lamp that falls directly on the floor for different room indexes and types of diffuser. It is the room index that defines the relationship between the height, length and width of a room. The utilization factor will be quoted in the product data specifications. The LDL or lamp design lumens will also be given in the product data specifications for the particular luminaire type. The maintenance factor has to be decided and in general this is 0.8 but can be lower if conditions are dirty.

The spacing tables to meet current UK and European Standards plus drafts can readily be applied to any application because the photometric design data is registered within the ICEL photometric scheme which requires that:

- the fitting complies with BS EN 60598.2.22;
- the manufacturer has ISO 9000 approval;
- the fittings are tested photometrically by the British Standards Institute;
- the derived spacing tables are subject to ICEL's third party inspection procedures.

The open (anti-panic) core areas can be separately referred to in appropriate spacing tables as illustrated at Figure 3.6.

This part of the design asks for 0.5 lux minimum of the empty core area which excludes a border of 0.5 m of the perimeter of the area. It should be noted that the spacing tables for 0.5 lux are derated to cover the diagonal points that exist in the installation.

Proposals in the industry will always promote improved levels of uniformity to ensure that a safe exit can be made from the area in the event of mains failure.

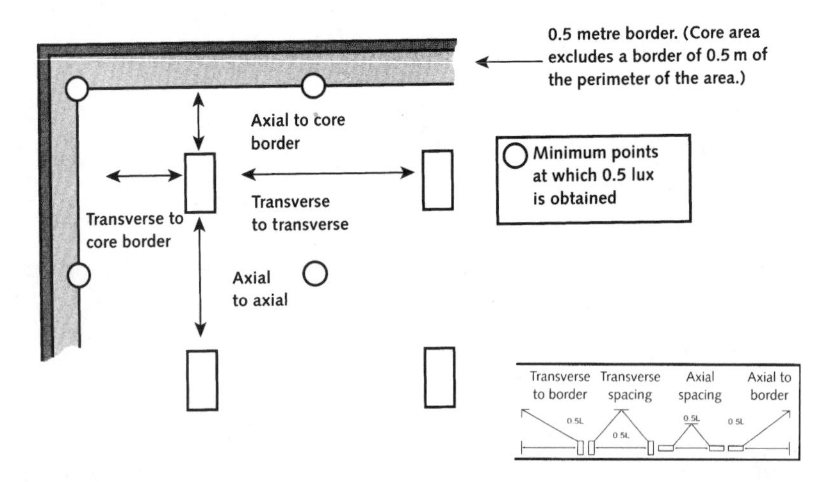

Figure 3.6 *Spacing – open (anti-panic) core areas*

3.6 High risk task area lighting

This sector of emergency lighting safeguards occupants involved in a potentially dangerous process or situation and allows the correct shutdown procedures to be implemented. This is for the safety of both operators and others using the building. The illumination is to be 10% of the normal levels but subject to a minimum of 15 lux.

Included in these areas are high physical risk environments of dangerous plant and production lines.

To design for these areas small fluorescent luminaires are not always satisfactory and it may be necessary to use a conversion of a normal luminaire or to employ a self-contained luminaire with two high intensity floodlight heads. To avoid undue glare these are mounted at least 30° above the line of sight. These units are particularly effective in premises with high ceilings and are ideal in the protection of hazardous areas with increased levels of illumination or to allow specific functions to be carried out. If it is necessary to employ high light outputs and the glare problem cannot be easily overcome a large lamp fluorescent with a conversion unit should be considered.

These units will be found to be of either 1 or 3 hours' emergency duration and have either square flood beams of 30° or have a pencil beam.

Spacings for these can be checked against the manufacturers' photometric tables.

To boost illumination in one area or to provide a general flood of light in an open area these should be aimed at some 90° separation. For long narrow spaces such as escape routes and warehouse aisles they are best aimed in opposite (180°) directions in order to obtain the maximum spacings.

When designing for these areas the illumination is to be 10% of the normal illuminance or a minimum of 15 lux whichever is the greater, and it is to be the task area that is illuminated rather than the floor. In practice the duration may need only be short term.

3.7 Exit signs – format and size

Exit sign luminaires are intended to direct and illuminate the means of escape as required by the Fire Precautions Act 1971 in compliance with BS 5266 and European requirements. The Health and Safety (Safety Signs and Signals) Regulations 1996 requires adequate provision of signs protected by emergency lighting. It specifies that signs are to be installed at all final exits and also on the escape routes at any location where the route may be in doubt.

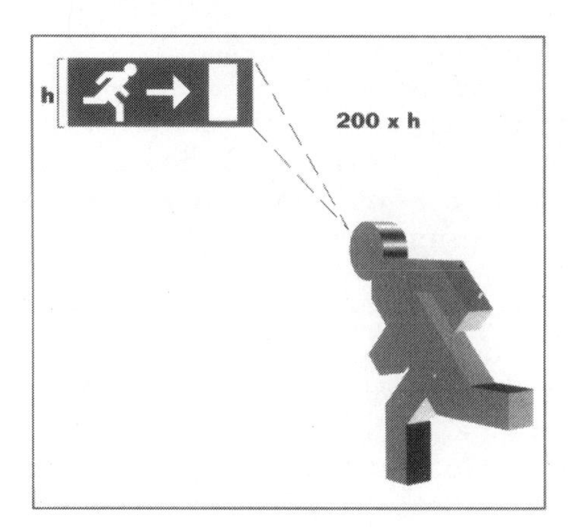

Figure 3.7 *Exit signs – maximum viewing distances*

BS 5266 states that exit signs are to be sited on all escape routes so that the occupants have no doubt where to proceed towards the final exit but this often means that exit signs are to be visible from any point on an escape route. To this end signs are required at all exits, emergency exits and escape routes, such that the position of any exit or route to it is easily recognized and followed in an emergency. Where direct sight of an exit or emergency exit is not possible and doubt may exist as to its position, a directional sign (or series of signs) should be provided placed so that a person moving towards it will be progressed towards an exit or emergency exit. It is from this that we recognize the need for exit signs to be visible from all points on an escape route.

Exit signs are to be wall mounted at a height of 2 to 2.5 m above the floor level. This height is that at which a human is accustomed to look for a sign. In Chapter 8 we consider low mounted way guidance systems and appropriate signs but these systems should be complemented with conventional signs at the stated height. Ceiling mounted signs are not recommended because legibility is inferior.

The maximum viewing distances are given as 200 × panel height for internal self-illuminated signs as shown in Figure 3.7. This can be reduced to 100 × height of the sign for externally illuminated signs. In the first instance the size and form of exit signs must be verified to meet the requirements of the standards in force plus health and safety guidance so they must be considered in relation to the maximum viewing distance.

The actual format of the sign must also be standardized and these are to be as shown in Figure 3.8.

Figure 3.8 *Exit sign format*

The European Signs Directive has introduced changes to the appearance of the familiar exit signs that originally conformed to the interim format within BS 5499. This is now obsolete and only acceptable if they are already in use in a building. The new pictogram sign with no wording came into force in April 1996 and was covered by Statutory Instrument 1996 No. 341. Descriptive wording in the Standard such as 'Fire Exit' is not used in the Safety Signs Directive and if it is required a separate sign-board should be provided.

It is important not to mix legends within a building. The new Eurolegends are based on the primary colour of green as this denotes a safe exit route together with a pictogram of a running person. This figure is displayed in two forms, facing left for a left exit or facing right for all other applications namely right exit, straight on, up or down.

It is to be noted that if the normal lighting is dimmed such as that in use in lecture theatres and cinemas the exit signs must be permanently illuminated using maintained lighting.

Self-adhesive legend panels with a standard colour of white characters on a green background can be affixed to most luminaires in order that they operate as exit signs.

3.8 Disability glare

Although we are obliged to site luminaires in accordance with BS 5266 further attention must be made to the manner in which they are positioned otherwise the installation may introduce disability glare which will obstruct the viewing of obstructions and signs. This can create a hazard in causing persons to stumble or collide with an object that would become obscured.

Glare is produced in those situations in which there is excessive luminance contrast and the presence of such a high luminance source causes an unfavourable reaction with the human vision in two main ways:

- The eye becomes less able to discern detail in the vicinity of the light source.
- There is discomfort and an effect of mild brightness distracts attention and leads to a sensation of light pain.

We can class glare as an excessive variation in luminance within the field of vision. It tends to be classed as either one of two groups: disability glare or discomfort glare. It may be considered as direct when it occurs as a consequence of bright light directly in one's line of vision. It may otherwise be recognized when light is reflected, for example, off highly polished surfaces.

The form of glare, namely disability glare, that we are concerned with is the form of glare defined as disabling an individual and creating a problem for them to carry out a particular visual task. Any individual subject to disability glare for a prolonged exposure time will experience a feeling of discomfort.

On a horizontal escape route, open (anti-panic) and high risk areas the luminous intensity in candelas (cd) of the luminaire is not to exceed that of the value in Table 3.1. This is within the zones 60° to 90° from the downward vertical in all angles.

Disability glare is referred to in angles because we need to look up further to be affected by glare, so that on a flat floor the glare zone is termed as 60° from the vertical. The zone is governed by a limit of

Table 3.1 *Limits of glare zone*

Mounting height (m)	*Limits of glare zone*	
	Escape route. Open area. Max. luminous intensity (cd)	*High risk task area. Max. luminous intensity (cd)*
<2.5 m	500	1000
>2.5 m <3.0 m	900	1800
>3.0 m <3.5 m	1600	3200
>3.5 m <4.0 m	2500	5000
>4.0 m <4.5 m	3500	7000
>4.5 m	5000	10 000

light intensity because as the height above the floor becomes greater the limit is increased also. However, in an area with hazards including steps the angle of vision varies so the glare zone must be regarded at every angle.

When considering high risk task areas it is necessary to pay particular attention to the glare zone since the illuminance is higher so the eye is more tolerant of the glare.

It can be noted that the widely accepted 8 watt fluorescent used in many emergency lighting systems cannot approach the glare limits, whereas projectors readily exceed the limits and are to be mounted as high as possible and should be directed down at an angle of less than 60°.

Figure 3.9 shows the concept of disability glare limits.

When concluding the design considerations and subject of disability glare some thought must also go into the response times of the luminaires and particularly if discharge lamps have been specified. The requirement for escape routes and open areas is that all of the lighting should operate within 5 s at an output of greater than 50% and within 60 s at full output. For high risk task areas the full illuminance is to be provided within 0.5 s.

High pressure discharge lamps have a warm-up phase and if used a no-break supply must be installed. These may also have problems with restrikes so delay circuits are needed to hold the emergency lighting on after the mains is restored and until the discharge lighting is operational.

For simplicity any discharge lamps are best kept as a separate identity to the dedicated emergency lighting.

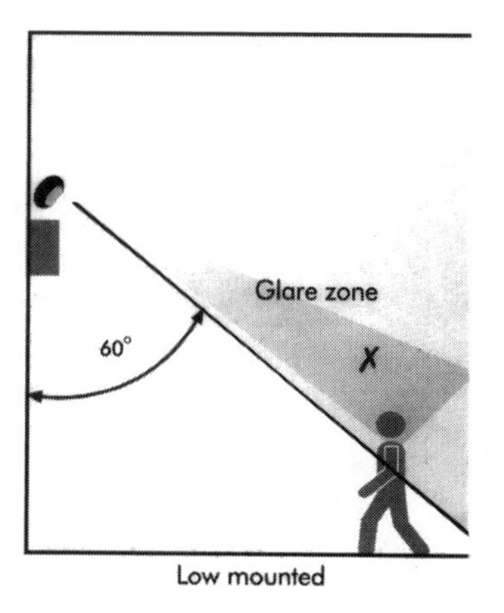

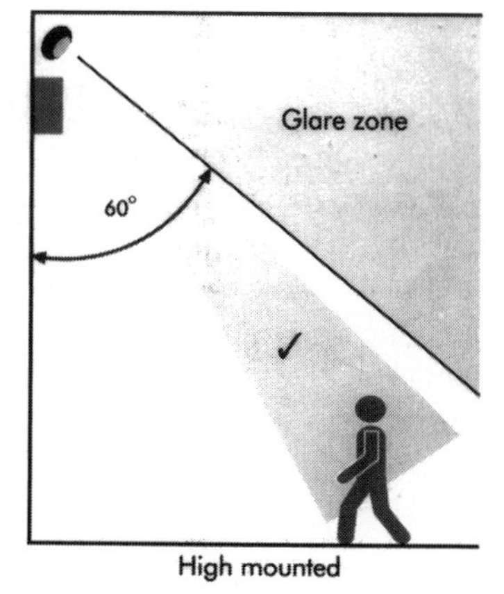

Figure 3.9 *Disability glare*

4 Self-contained systems

These systems have luminaires with an integral charging system, battery, control gear and lamp. In the event of mains failure the lamp will be operated for the design duration via the onboard battery. They also incorporate a mains failure relay/semiconductor system and charging indicator LED and may have an inverter for operating a fluorescent tube. These will be found to incorporate control gear to operate tungsten lamps with outputs from 2.4 watts, miniature fluorescent tubes and fluorescent tubes with capacities up to 80 watts. In exceptional circumstances even greater ratings can be attained.

They are called self-contained because they contain all of the system components together with the battery and lamp within one unit. Self-contained systems are easily understood and installed and are used extensively in small buildings.

Self-contained luminaires consist of special design automatic charging systems in conjunction with rechargeable sealed battery cells of nickel–cadmium type. These are normally suitable for continuous use at a case temperature of 50°C. Self-contained luminaires need a permanent mains connection from the same fuse as the general lighting for that area. Maintained and combined luminaires can have an additional separate switched feed in order that the luminaire can also be switched in the conventional fashion.

Continuity of the battery charging circuit and the presence of the mains supply are monitored by an LED. There is little maintenance needed and minimal attention to the unit is required other than routine checks and cleaning.

The advantages and disadvantages of the self-contained luminaire system are:

Advantages

- Installations are fast and inexpensive. In addition they are easily expanded and adapted as the requirements of a building change.

- The wiring is in the same cable type as used for the other areas of the building. In general this is PVC insulated and sheathed. If the other wiring in the building or in the same area was to be destroyed this would bring on the local emergency lighting.
- The wiring is easy and should be installed in such a way that failure of the local protective device feeding the lighting cables will activate the emergency lighting for that area.
- Failure of one luminaire will not create a problem for others as all circuits are in parallel.
- There is inherent proof against the supply failure as the luminaire has an integral battery.
- Maintenance is minimal.

Disadvantages

- The luminaires are relatively expensive since they all contain a battery and control gear.
- The batteries have a limited life and need to be replaced after a 4 year working life period.
- Systems are susceptible to working in high ambient temperatures.
- Luminaires for maintained systems have onerous temperature limitations that can further affect the onboard batteries.

Self-contained luminaires operate in non-maintained, maintained or combined (sustained) operational modes as shown in Figure 4.1.

Non-maintained (NM). In this mode the emergency luminaire is off when the mains supply to it is healthy. In the event that the supply to it is no longer available the lamp is illuminated from the units integral battery. Once the mains supply is restored to the luminaire the lamp will switch off and the battery will recharge to its full potential.

Sustained luminaires have additional mains circuitry and from a legislation point of view are regarded as non-maintained.

Maintained (M). This mode is similar to the non-maintained version but the lamp can also be used when the mains supply is healthy by means of a switched live connection. In order for this function an additional connection is made. In the event that the mains fails the luminaire operates in the same fashion as the non-maintained luminaire by the lamp drawing its power source from the integral battery.

Combined. This may also be called a sustained luminaire and is a unit containing two or more lamps at least one of which is energized from

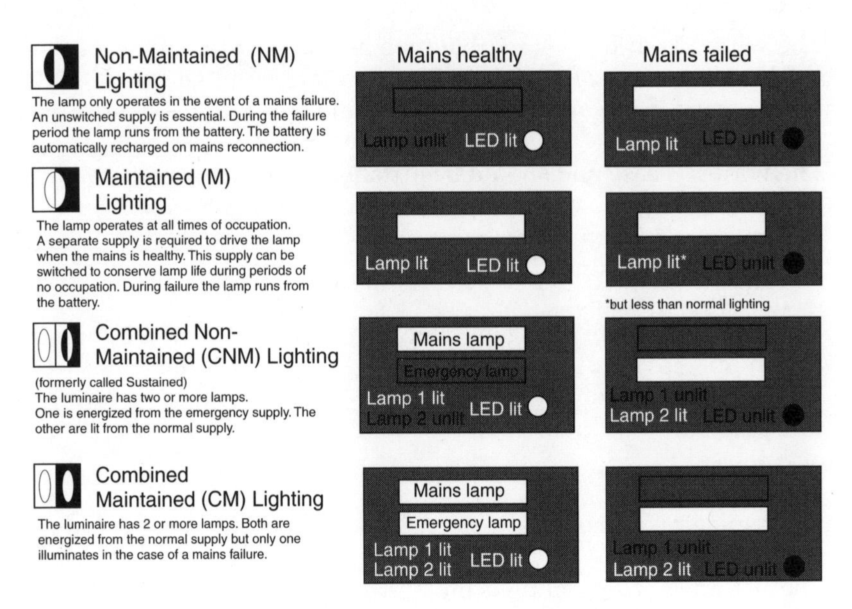

Figure 4.1 *Self-contained luminaires*

the emergency supply with the remainder energized from the normal supply.

With these units if the emergency lamp is only illuminated in the event of a failure of the mains the luminaire is regarded as non-maintained.

There is also a combined maintained lighting variant which is also a luminaire containing two or more lamps. Both of these are energized from the normal supply but only one illuminates in the event of a mains failure.

The wiring of the self-contained luminaire is shown in Figure 4.2.

4.1 Wiring techniques

The wiring of self-contained luminaires is effectively with the loads being in parallel circuits.

In all cases the wiring of self-contained luminaires is to be such that it prevents unauthorized disconnection. However, there should be a facility incorporated so that a mains failure can be simulated for testing purposes.

It is vital that the supply is from the same local fuse as the normal lighting so that if a failure of the local fuse occurs, either because of

Non-maintained installation

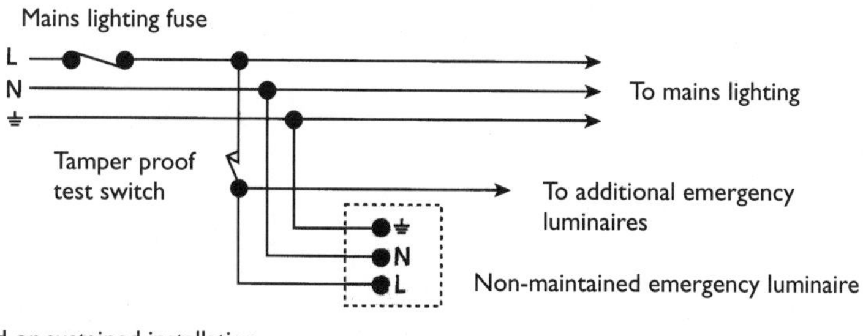

Maintained or sustained installation (fluorescent)

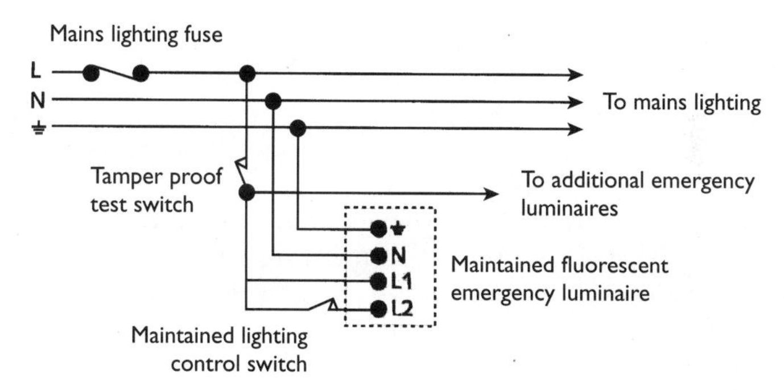

Figure 4.2 *Wiring – self-contained luminaires*

a fault, or because it has been deliberately induced by an act of vandalism to cause disruption, the emergency lighting will become effective for that area. If the local fuse is not used the emergency lighting would only illuminate if a total power failure was to occur.

The wiring of the luminaires should be in accordance with the IEE Wiring Regulations and be unswitched from the local lighting source. If a switched live is also added to provide the maintained feature it must also be derived from the same origin.

Maintained and combined non-maintained (sustained) luminaires are always to be wired with a separate switched feed as it is useful to be able to operate any maintained light in a switched circuit that can be turned off at night or when the building is unoccupied. A maintained light that is permanently energized will have its working life reduced. The cabling that is installed for the emergency lighting should be the same as that in use for the normal lighting circuits. If the cabling for the emergency lighting was to have additional protection it could inhibit the emergency lighting from operating even though the normal lighting circuits had been destroyed by fire. For this reason the cabling should be PVC insulated and sheathed.

Before considering the typical self-contained luminaires in common use and their performance we can overview the self-contained mode of operation:

- Non-maintained luminaires only come on when the normal supply fails.
- Maintained versions remain illuminated under all conditions.
- Combined or sustained luminaires have a particular lamp illuminated when the mains is normal and a different lamp on when the supply is disrupted.

The batteries used in this system tend to be high temperature sealed nickel–cadmium or sealed lead–acid and are an integral part of the assembly. The actual battery capacity is the discharge capability of a battery, being a product of average current and time expressed as ampere hours over a stated duration. It may be noted that a shorter total discharge period gives rise to a smaller available capacity. The emergency duration is the time that the battery will power the lamp in the event of a mains failure. The emergency lamp output is the light output quoted in lumens and the normal lamp output is the light output in lumens with the mains healthy. The light source tends to be tungsten working direct from the battery or fluorescent that is powered from an inverter with the battery at the source.

The rated duration is the manufacturers' declared duration for a battery operated emergency lighting unit, specifying the time for which it will operate after mains failure. This may be for any reasonable period, but is normally 1 or 3 hours when fully charged. The duration should normally be 3 hours, but shorter durations are sometimes permitted. Licensed and residential premises must have 3 hours, but in industrial premises 1 hour is generally acceptable. A list of permitted durations is tabled in ICEL 1003. The recharge period is the time necessary for the batteries to regain sufficient capacity to enable the lamp to perform its rated duration.

The use of maintained or non-maintained systems is often specified by the local authority. It is possible to generalize and non-maintained systems are incorporated in buildings with a much more limited occupation time. This includes offices and shops. A maintained system, however, is at an advantage in that in the event of a mains failure affecting the local subcircuit the lighting will already be on so that local subcircuit monitoring is not necessary. Maintained lighting is also useful for low level night lighting and for keeping premises lit for security purposes so that the area is always illuminated as a first defence against intrusion. In places of entertainment and licensed premises maintained systems are normally specified.

As a conclusion to this section we can overview the emergency duration:

- One hour is the minimum required of any application.
- A 3 hour duration is needed for entertainment, sleeping risk, if the evacuation need not be immediate or if early reoccupation following a short mains failure is required.

There are two control modes, namely inhibit mode and rest mode, that relate to and have an effect on the discharge of the emergency batteries with respect to the duration. These are defined in Chapter 10, section 10.1.

4.2 Luminaires – range of variants

It is only possible to give an overview of varieties available because these do differ in specification between the competing manufacturers and their performance criteria.

There is a huge range of self-contained luminaires available to meet any application and these will extend from bulkhead fittings to stylish circular versions with crystal diffusers. Domestic applications may even be satisfied by employing a design of particular aesthetic appeal.

For more demanding applications many units are weatherproof to high classes such as IP 65 together with good levels of resistance to vandalism, and flameproof versions can also be purchased for use in hazardous areas where flammable gases and substances can be encountered under extreme conditions.

The most popular luminaire for mainstream emergency lighting is the 8 watt bulkhead.

It will have a typical specification as follows:

- Heavy duty die cast aluminium body with cast cable channels and an epoxy powder finish, or with the body manufactured from ABS.
- Having a high impact opal polycarbonate diffuser with an oil- and waterproof neoprene gasket. Weatherproof to IP 65 with the mounting holes outside of the sealed area.

- Steel gear tray for ease of installation.
- Quick release terminal blocks to isolate the chassis and facilitate in and out wiring.
- Optional semi-recessed flange.
- Fully constant and automatic current charging system.
- Maintenance-free sealed nickel–cadmium batteries with a high temperature polypropylene separator.
- Certified to BS 4533.102.22 and ICEL 1001.
- Highly visible mains charging monitoring LED.
- High frequency high efficiency inverter.

System mode	non-maintained or maintained
Emergency duration	3 hours
Light source	300 mm 8 W fluorescent tube
Emergency lamp output	180 lumens
Normal light output	180 lumens
Battery	high temperature nickel–cadmium 3.6 V 4 Ah
Recharge period	24 hours (or 14 hours for 1 hour duration)
Charging current	275 mA
Charging monitor	LED
Input voltage	230 + / – 10% 50/60 Hz
Cable entry	BESA in rear, surface from sides and ends
Power consumption	8 VA non-maintained/20 VA maintained
Environment	0 to 25°C
Weight	1.1 kg non-maintained/1.3 kg maintained
Dimensions	As Figure 4.3

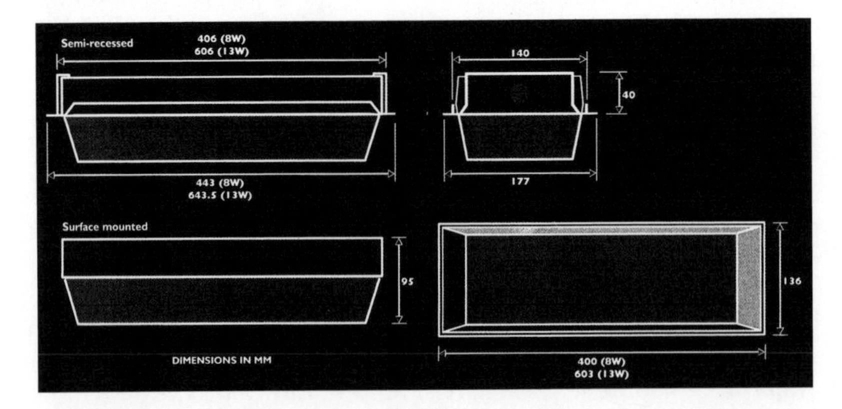

Figure 4.3 *Self-contained bulkhead – essential dimensions*

Although smaller luminaires using 6 W fluorescent tubes are available in either bulkhead or square form housings their emergency lamp output is more in the order of 100 lumens. These tend to be used to enhance décor and where physical dimensions are of increased importance.

Although 8 W and 6 W fluorescent tube luminaires remain popular and particularly the former version, stylish circular luminaires also have a role to play where aesthetics are important. They meet the requirements for discreet fittings for prestige applications.

These tend to have spun steel bases and crystal diffusers with bayonet fixing glass retention so that the glass is held if it works loose. Luminaires of this type have single or two 4 W emergency lamps with 60 and 90 emergency lamp lumens respectively. The units can be surface or recessed mounted.

Further decorative versions for semi-recessed mounting have 2.4 W krypton filament lamps. In these cases they have some 22 emergency lamp lumens.

Downlighters with simple spring clip fixing methods reduce installation time on site. These require a minimum of wiring and because they are of self-contained form have no remote battery box to be fixed and wired. The recessed position of the lamp cuts off the light at the optimum angle so avoiding disability glare whilst giving a good light distribution allowing excellent spacings for 1 lux at a mounting height of 3 m. By the use of 4 W tungsten halogen bulbs these units are specified at emitting some 53 emergency lamp lumens.

Decorative luminaires to suit a variety of locations from hotels and shops to offices and colleges have a white or light grey fire retardant polycarbonate body and complementary opal diffuser. These are of a circular or square low profile shape allowing the unit to blend in with most interiors and furnishings. The emergency self-contained luminaires in this range use high efficiency 4 W fluorescent lamp or 10, 16 or 28 W 2D fluorescent lamps.

Emergency self-contained 3 hour duration units will have a typical emergency lamp output as follows:

Emergency lamp	*Emergency lamp lumens*	*Luminaire*
4 W fluorescent NM and M	60	215 mm diameter
10 W 2D fluorescent NM and M	210	215 mm diameter
16 W 2D fluorescent NM and M	190	285 mm diameter
28 W 2D fluorescent NM and M	220	260 mm square

Self-contained high intensity twin lamp units are most suited to high bay industrial use, long corridors, warehouses and supermarkets. These are a powerful robust system comprising two adjustable floodlights each on a knuckle joint mounted on top or to the sides of the charger battery case. The floodlights are brought into operation by an automatic mains failure relay. These will have a rating in the order of 12 V, 12 W or 20 W using tungsten halogen lamps giving an effective beam of up to 28 m.

The case which houses the battery and charging unit is of stove enamelled sheet steel or fire retardant noryl plastic with a mounting bracket at the rear of the enclosure which can be detached from its mounting and fitted to the wall for easy installation. In these units the battery is a 12 V sealed lead–acid type with protection against deep discharge by means of a low voltage disconnect system. Protection is afforded by fuses and a test switch is added together with a charge indicator light to prove battery and charge continuity.

These luminaires can be equipped with a delay timer when it is for use with backup discharge lighting which has a relatively slow warm-up time. Remote lampheads can also be supplied to accommodate mounting at an additional remote point.

It will be found that many of the products offered by the main manufacturers can be retrofitted but there will always be specialist luminaires for more diverse applications. For example, flameproof luminaires have been developed for use in hazardous areas where there is a probability that explosive gases could become present in the atmosphere and form an explosive mixture. Typical applications are refineries, oil rigs, oil tankers and chemical plants, and pharmaceutical processing areas. These luminaires are of a special construction and have particular approvals to cover their duty in clearly defined environments.

Although exit signs legends are available for application to many emergency luminaires other luminaires are designed for use specifically with these signs. Such luminaires are supplied complete with a self-adhesive legend kit enabling a wide variety of pictogram style signs complying with the European Signs Directive to be affixed to luminaires. These are ordered as legend panels specifically for the luminaire to which it is to be attached. Other legends can also be supplied by manufacturers to satisfy special customer orders.

From this we can say that self-adhesive legends can be affixed to many luminaires to show the escape route. However, edge illuminated signs are actually luminaires in their own right and are covered in the next section.

4.3 Edge illuminated signs

Exit sign luminaires are designed to both illuminate and direct persons to the 'means of escape' as invoked by the Fire Precautions Act 1971 and in compliance with BS 5266 Part 1 and European requirements. They are available with a wide range of fittings as shown in Figure 4.4.

Edge illuminated signs tend to be of the following type:

- *Wall mounted*. These are fitted onto a wall mounted support termination box or suspended from it by chains.
- *Ceiling hung*. The luminaires are hung by chains from a ceiling mounted support and termination box. The chains are adjusted for height. It is otherwise known as a cantilever bracket.
- *Swan neck*. Intended for spacing away from a wall.
- *Cantilever bracket*. These are suspended by chains attached to a bracket which is of itself attached to the wall.
- *Recessed*. In this case the fixing kit is recessed into the ceiling void so that only the legend panel is visible in the premises.

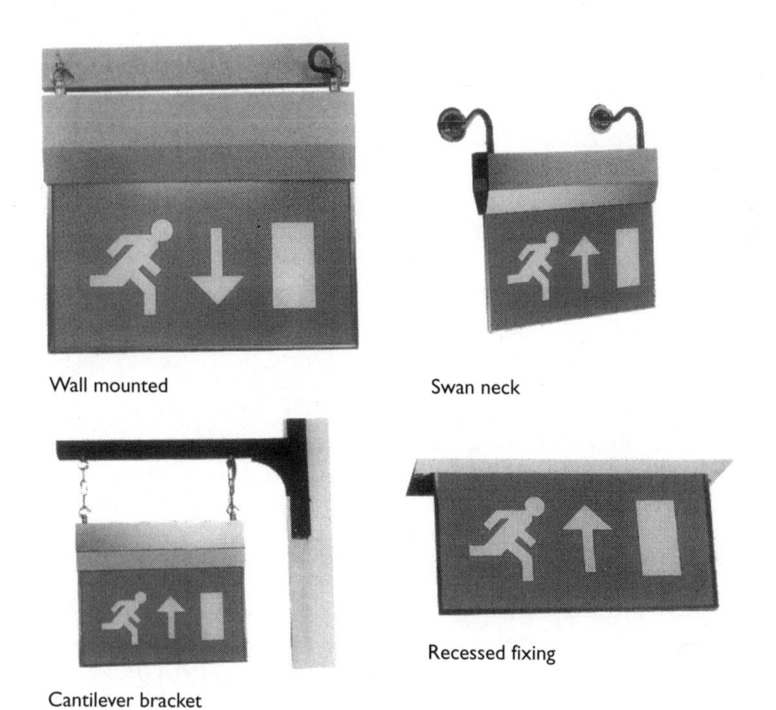

Figure 4.4 *Edge illuminated signs*

All of the noted edge illuminated signs must be given an even spread of light by the illumination source that is concealed in the main housing internally lighting the high visibility message.

These signs add distinction in public areas where clear direction, together with good appearance, are important. All versions tend to be illuminated by 300 mm 8 W fluorescent tubes in maintained or non-maintained mode.

Exit downlighters are often found where higher levels of illumination are needed on escape routes in accordance with European Regulations. These are effectively a light controller comprising a robust enclosure with a slide-in protected panel to resist tampering and vandalism, and come complete with the self-adhesive legend kit. They can achieve cost savings because they reduce the need for additional emergency lighting within 2 m of the exit sign by using a polycarbonate light controller giving economical operation and a good spread of light to the legend and the surrounding area.

It cannot be too highly stressed that the luminaires available to us can satisfy any application since there is a range of highly decorative units together with good illumination properties. The installation can be further enhanced by the use of good quality edge illuminated signs.

As a conclusion to self-contained systems there is a need to understand the subjects of fault finding, commissioning and testing. These have relevance also to the signs that are a part of the installation and form the next section.

4.4 Fault finding – commissioning and testing

Fault finding

Following the installation of a new self-contained system or the replacement of an existing luminaire, if any fault condition exists there will be a need to take corrective action if the testing is not satisfactory.

Indicator LED not illuminated:

- Ac supply interrupted and should be restored.
- Unit track fuse blown and should be replaced with an axial fuse. (Fuse value is marked on the printed circuit board.)

Unit does not meet required emergency duration period:

- Ensure that the luminaire is not operating outside of the unit's temperature limits.
- It is possible that the unit may need recycling and should be discharged and recharged for 24 hours. The unit should then be retested and if the emergency duration has improved the discharge/recharge is to be repeated a further time.
- Battery pack may be in need of replacement. These are readily available.

Lamp is not fully illuminated:

- If this is found in mains failure the tube ends should be checked to see if blackened and if so a new replacement tube should be installed.

No illumination:

- If this is found in mains failure it is necessary to ensure that the supply is disconnected before disturbing the wiring. The wiring is then to be checked visually for connections with corrections carried out as appropriate.

Commissioning and testing

Once the luminaire has been installed a 24 hour charge cycle should be performed. After this period the LED indicator should be checked to ensure that the batteries are on charge.

A short discharge test should be carried out to check correct performance of the luminaire by isolating the unswitched supply. After restoration of the supply the luminaire should be checked a further time to confirm that the LED indicator is illuminated.

The following procedure is intended to ensure the continued protection of the premises and occupants.

Because of the possibility of a failure of the normal lighting supply occurring shortly after a capacity self-test of the emergency luminaire, or during the subsequent recharge period, all tests should wherever possible be undertaken at a time of minimum risk. It is important that the emergency luminaires are commissioned alternately or in groups that minimize the risks as outlined. After any test or the restoring of the mains supply it is a requirement to check that the LED is illuminated.

A test procedure is to be implemented in accordance with BS 5266. In summary the recommendations are:

- *Daily test.* Visual inspection of the LED indicator to ensure that it is healthy. Ensure that maintained luminaire lamps are lit.
- *Monthly.* Check system by testing for a short period. This need not exceed 25% of the rated duration. Often this is performed as an operational test of 30 seconds.
- *Six monthly.* Test for a 1 hour period, for a 3 hour system. This is to effectively carry out a test of one-third of the rated duration of the luminaires.
- *Three yearly.* Test for the full rated duration and subsequently test annually.
- *Every subsequent year.* A repeat of the 3 year test.

During each test, the lights are to be checked for correct operation. After testing the supply should be restored and the indicator lamps or meters checked to ensure that the system is charging correctly. On completion of the commissioning the drawings of the installation should be retained on the premises and kept up to date as necessary. It is also a requirement that a copy of the luminaires' operational status and description of the performance be retained in the site log book.

In order to prove the efficiency of the full self-contained system a test circuit that may be installed to correctly switch the mains supply can be applied as shown in Figure 4.5.

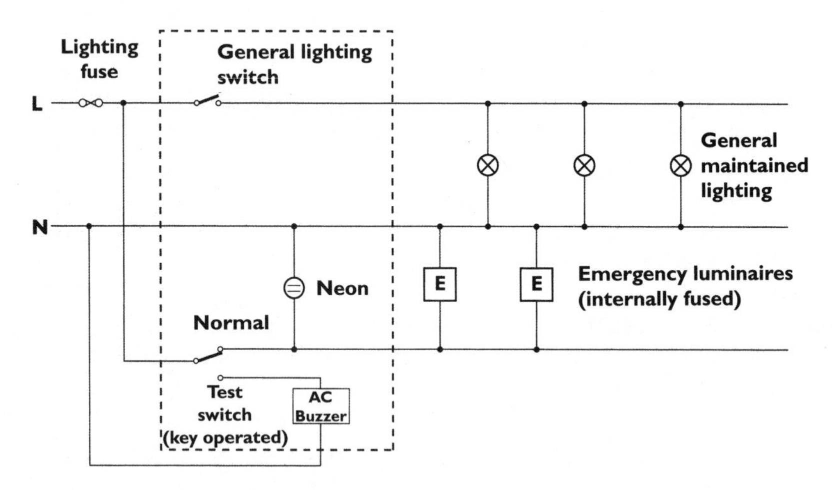

Figure 4.5 *Test circuit – self-contained luminaires*

TEST RECORD SHEET

Luminaire location: Installer: Engineer: Date:

Product Type:	Mode:	Duration:	Full re-charge time:	Lamp size:

Month	Test	YEAR 1		YEAR 2		YEAR 3		YEAR 4		YEAR 5	
		Signature	Date	Signature	Date	Signature	Date	Signature	Date	Signature	Date
1	Functional										
2	Functional										
3	Functional										
4	Functional										
5	Functional										
6	1 Hr Duration										
7	Functional										
8	Functional										
9	Functional										
10	Functional										
11	Functional										
12	1 Hr Duration										
	3Hr Duration										

Figure 4.6 *Test record sheet*

The test circuit as shown ensures that the emergency circuit can be tested without a need to disconnect the general lighting. The use of a buzzer gives audible warning that the key switch is in the testing position so that the system can not be left in the discharge mode by accident. The assembly can be made up in a grid box to include the general lighting switch, the test switch, the buzzer and a neon which is effectively wired across the load of the internally fused emergency luminaires.

It may be noted that the emergency lights are wired in a parallel circuit looped in and out so that the test switch switches the loop at source.

The buzzer in the illustration is a mains ac version but a 12 Vdc type could be used by means of switching a 240 V relay coil and driving the buzzer via the switched contacts with power derived from a low voltage power supply. Following all of the testing a Test Record Sheet should be completed typically as shown in Figure 4.6.

5 Central battery systems

The central battery system is a type in which the batteries for a number of luminaires are housed within one location. This may be for all of the luminaires in a complete building or more usually for all of the luminaires on one lighting subcircuit. In practice therefore the batteries that power the system in the event of a mains failure are, unlike the self-contained system, held remote from the luminaires.

In accordance with the European requirements these types may be of the small central system, large central battery system or central inverter system.

It is normal that central battery systems are engineered to satisfy a 3 hour emergency duration although in certain industrial applications a 1 hour duration can be accepted.

There are a number of modes that central battery systems can operate in. Refer to Figure 5.1.

- *Non-maintained*. Load supplied only during mains failure conditions. In this case the output is dc. The battery is switched to the load only in the event of a failure of the mains supply. This is performed by a contactor.
- *Maintained*. When the mains is healthy a maintaining transformer provides an ac supply to the load. In the event of the mains failing the load is supplied as a dc supply by the battery which is connected by the switching over of an automatic contactor.

  It is possible to switch the maintained output both on and off using a maintained switch that can be sited on the cubicle holding the control gear or by a switch sited remotely and connected to the remote terminals. These switches may be called night watchman's switches. If the mains fails the battery is always switched to provide the load irrespective of the position of any maintained switch.
- *Floating output*. The load to the luminaires is from a permanently connected battery charger system giving a dc source. These outputs can be either standby or continuous load. The charger must be rated so that it is able to supply both the required battery current and the system load current.

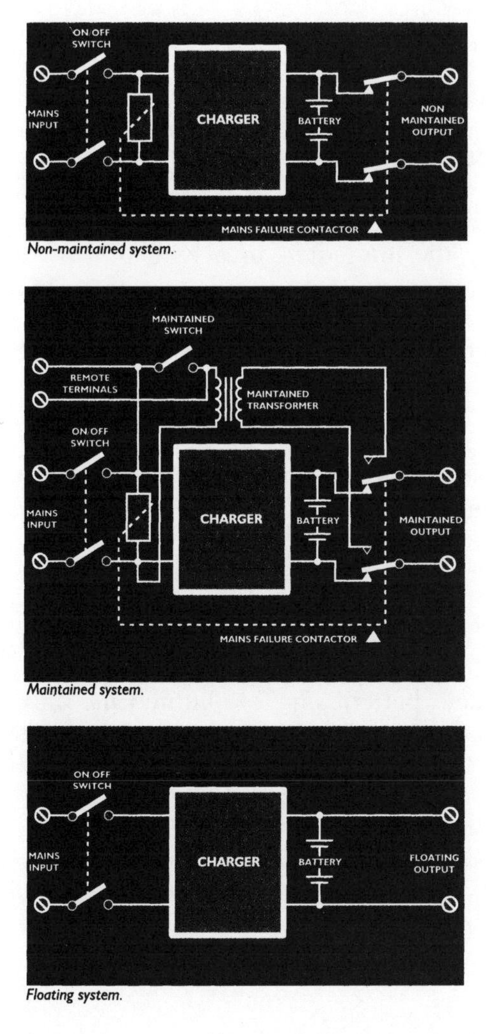

Non-maintained system.

Maintained system.

Floating system.

Figure 5.1 *Modes of operation – central battery systems*

It will be apparent that because the power for any emergency critical system in which the energy source is at a remote point from the luminaire there is a need for special cable to afford protection against fire or associated damage. In addition consideration must always be given to voltage drop between the battery and luminaire. In order to keep the voltage drop down to an acceptable level there is a need to use relatively heavy cables in relation to many other electrical engineering techniques. It may be noted that BS 5266 specifies that the voltage drop is not to exceed 10% of the nominal. However, in many applications it is prudent to design for a drop more in the order of 5%

to cater for temperature variations, termination contact resistance and any future expansion of the network.

5.1 Small central systems

These are compact versions employing sealed lead–acid batteries in order to supply the immediate area with power for the emergency lighting. The outputs will be in the range of up to 24 V 408 W for 3 hours or 720 W for a 1 hour duration.

They have certain advantages and disadvantages:

Advantages

- The distribution costs are reduced.
- The cost of the equipment is low.
- Little maintenance is needed.
- It is easy to provide local mains failure protection.

Disadvantages

- Sealed lead–acid batteries have a limited life and in general need replacing between 4 and 5 years after commissioning.
- The batteries lose capacity with age and are only 80% efficient after four years.
- The battery life degrades above 45°C.
- Special cable is needed unlike that used in self-contained systems. MICC cable is used extensively as it offers high levels of protection to fire and impact.
- There are a number of systems to test unlike a single large central battery system.

Systems of this nature are intended as budget cost with low maintenance of the batteries being needed because they have a constant voltage charging system. They are most suited to small installations such as restaurants and public houses because they require little attention.

A robust cabinet holds all of the system control gear and the batteries. The control equipment will typically have as standard:

- Mains switch/MCB and neon.
- Maintained switch/MCB and neon.
- Charge failure alarm.
- Mains failure alarm.

- Low volts disconnect system components to isolate and protect the batteries from deep discharge as a result of an extended mains failure or cyclic power cuts.

Options will also exist to include:

- Charge ammeter to show the battery charge current.
- Battery voltmeter to show the battery volts.
- Additional maintained outputs.
- Additional non-maintained outputs.
- Time clocks (24 hour/solar/programmable) for automatic control of the maintained outputs with a specified battery reserve.
- Phase failure monitor to detect failure of separate phases.
- Fire alarm monitor to operate all emergency circuits via a 240 Vac relay coil to activate the auxiliary contacts of a fire alarm control panel.
- Subcircuit failure to operate all emergency circuits in the event of any local subcircuit failure.
- Subcircuit monitor using a relay to control a specified output on local mains failure for maintained or floating systems.

The luminaires for use with these systems are known as slave or centrally supplied luminaires. They are defined as an emergency luminaire without its own batteries that is designed to be used in conjunction with a central battery system.

An extensive range of luminaires and signs is available for use on either ac or dc supplies. These luminaires extend from interior commercial, interior decorative, tungsten floodlights through to external weatherproof types and include dedicated exit signs. In practice these can complement mains luminaires and be used in schemes to match existing fittings. When ordering slave luminaires it is essential that the system voltage is specified. Slave luminaires can also be specified with mains failure or changeover mains failure relays.

Conversion modules

These are designed for installation inside a standard general lighting luminaire. This enables it to operate as both a normal light unit and as an emergency luminaire. These modules only come into operation when the local supply fails when used in conjunction with a maintained or floating output system. By using such conversion modules there can be a policy of selectively converting existing general lighting. The installation of the module requires modifications to the internal wiring of a luminaire and this is normally performed by the manufacturer of the equipment at their plant.

In cases where it is not possible to fit a module within the luminaire because of space restrictions or if the internal temperature would be excessive it can be installed in a small enclosure at a remote point but close to the luminaire and within a distance of 1 metre.

It should be understood that during normal conditions using the standard switched mains supply and the conventional control gear that the lamp will operate at full brightness. However, when the emergency conditions are encountered the luminaire will continue to operate but at reduced brightness as the emergency lamp will be powered from the conversion module rather than from the conventional control gear.

Almost any mains luminaire can be subject to conversion for emergency use by the manufacturer but in those cases in which more than one lamp is present in the luminaire it is normal to convert only one for emergency use.

In performing the conversion there is a need to comply with EMC requirements and to a system of assessed capability to ISO 9000 standards.

The documents to support the conversion should relate to light output, photometric data, emergency ballast temperature, battery and ambient temperatures together with the connections for the final installation.

5.2 Large central battery systems

These are effectively one large central system network unlike the small central system of which there may be several in one building. They comprise a large battery up to several hundred Ah connected to a sophisticated charging circuit that must maximize battery life and minimize the loss of electrolyte.

Such networks are used in medium to large emergency lighting installations and particularly when there is a requirement for a high volume of luminaires. If there is a need for central control and testing, or if the lighting is to be used in either high or low ambient temperatures these systems are at an advantage to self-contained luminaire networks.

The architecture is as shown in Figure 5.2. In practice the systems are best sectioned into local subcircuit groups. Each group can have a separate distribution board and when there is a local failure the emergency lights that are fitted in that area become energized from the maintained supply.

These systems can be so large that they need to be contained within a special designated room although the more compact system may be found in a cubicle that is suitable for wall mounting.

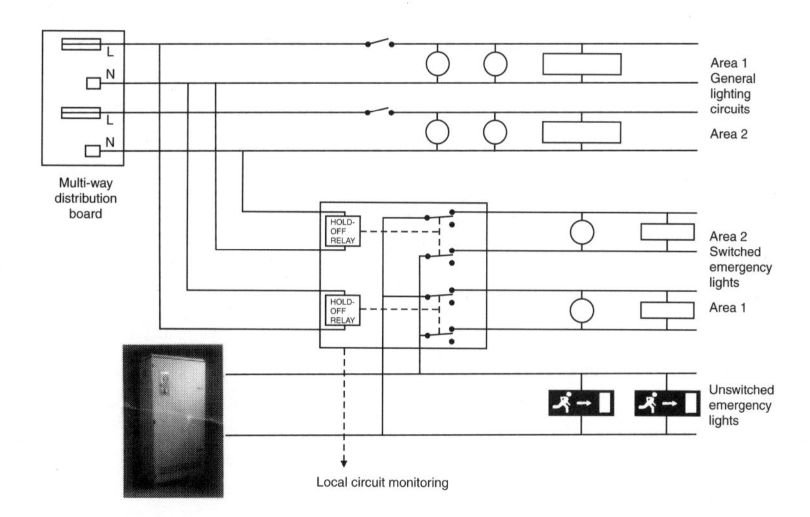

Figure 5.2 *Architecture – large central battery system*

As with any system network they do have certain advantages and disadvantages.

Advantages

- A good control of battery charging can be made because only one central unit is used.
- The system is of low cost and slave luminaires used in the system are inexpensive.
- The cost per watt of power is low.
- The batteries have a long life span in relation to small central and self-contained systems.
- It is possible to obtain a high light level for increased luminaire spacing distances.
- Luminaires can operate in relatively high or low ambient temperatures with a lesser degradation of the system.
- Testing is easy without interrupting the normal lighting.
- Testing and maintenance is easier to carry out than that of the self-contained type.
- Batteries are quickly and easily replaced.
- Central control is made over all of the slave units.
- A choice of emergency durations exist, either 1, 2 or 3 hours.

Disadvantages

- Installation costs are higher than those of the self-contained system.

- Segregation of the cabling is needed and it must be protected from mechanical or other forms of abuse or damage that could occur in the premises within which it is installed. It is also to be fire resistant.
- A battery failure will disable the whole system unlike the self-contained types which will only create an individual fault in the event of a battery failure.
- It is more complex arranging operation in the event of a local mains failure unless maintained systems are used; subcircuit monitoring may be needed.
- Certain battery types must be sited in a ventilated area.

There are many combinations of system types and modes in which the lighting can be used in conjunction with the normal lighting. An initial selection can be made depending on whether the lighting is to be integrated with the normal lighting by referring to the selection chart as shown in Figure 5.3.

Large central battery systems tend to be found in nominal voltages of 24, 50, 110 and 240 volts with the lower voltages the more economical. For large installations it is practice to use the higher voltages to retain more easily current levels and voltage drops within specified working limits.

The cabinets housing the batteries and control gear plus the instrumentation are in general formed from one cubicle only but certain large systems may be housed in multiple sets. It is normal that the cabinets are manufactured from heavy gauge 1.6 mm minimum steel sheet with a stove enamel finish and assembled to create a robust cubicle. The base of the cabinet is built on a plinth so as to prevent the accumulation of moisture and corrosive materials that could damage the cabinet paintwork.

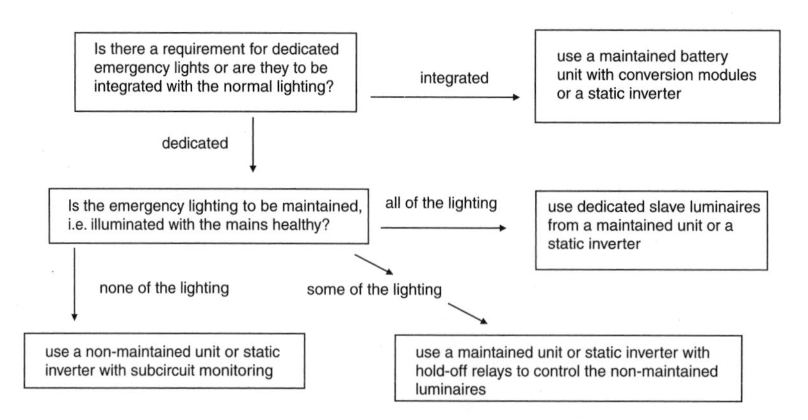

Figure 5.3 *System selection chart – large central battery systems*

The electrical control gear is in a separate apartment and segregated from the battery storage sections with different access doors. The battery compartments have tiered sections to enable the electrolyte levels to be easily checked. The cabinet has a fixed fascia panel to hold the controls and instruments. The controls and fascia panel of the charger compartment differ in layout between the different manufacturers but the typical display features are:

- Selectable digital. Voltage/charge/discharge.
- Current display. Green LED to indicate power on.
  Green LED to indicate float mode.
  Amber LED to indicate current limit.
  Red LED indicates either manual or boost mode selected.
  Green LED indicates maintained lights on.
- Fault LEDs to indicate battery low volts; battery high volts; charge fail; mains fail; + or – earth fail.

The central battery systems from the different manufacturers will have various extras but we often find options in systems such as dual output circuits with separate circuits on maintained modes for non-maintained and maintained lights and signs. Particular three phase monitors can automatically detect phase failure and energize all of the emergency lighting circuits and multiway subcircuit monitors can detect lighting circuit failure and energize the entire emergency lighting system. To achieve this monitoring relays are held within the cabinet and require a supply from each lighting circuit.

Integral HRC fuse distribution boards provide individual pairs of HRC fuses to feed the load output circuits for systems in which more than one circuit is required. As an option double pole MCBs can be used to protect the load output circuits. Timed test switches can be added to simulate a normal supply failure with automatic reset after a predetermined time. Fire alarm relays with a 24 Vdc coil can automatically energize the emergency lighting from the fire detection system. Time clocks for automatic timed switching of the maintained lighting can also be incorporated.

There will also be a range of remote optional extras for use in conjunction with central battery systems. Remote alarm units can provide a common, visual and audible indication of a system fault. These can come complete with a sounder mute facility. They automatically reset at the point in time that the fault clears.

Subcircuit monitors for non-load switching are used to monitor the status of the normal mains system lighting. These provide a signal to the central battery unit control gear in the event of failure of the final lighting circuit. If any of the monitored ways fails the switch contacts open. If a monitored supply fails a contactor drops out and allows the maintained supply to be connected to the emergency luminaires.

Hold-off relays for load switching are used to hold off the maintained output from the central battery unit. This provides non-maintained luminaire operation with monitoring of the normal final lighting circuit.

5.3 Central inverter systems

The object of a central inverter system is to supply mains power to the emergency lighting circuit. To this end the inverter is used to convert the central battery voltage to mains potential in the event that the normal lighting supply fails. It is this converted supply that is used to energize the mains luminaires.

The system incorporates control equipment to stabilize the mains output voltage to compensate for the battery voltage variation during the time of its discharge. The inverter uses pulse width modulation techniques to stabilize the output and adopts a filter to achieve a pure sine wave. However, it is to be understood that the output voltage and frequency of the waveform may differ from that of the actual mains supply.

We can therefore see the inverter as a device for operating more than two lamps to energize remotely sited mains luminaires.

The central inverter system uses a static inverter battery cabinet and integral control gear to convert dc battery power to ac at an appropriate voltage and frequency. The main advantage of this is that it can use standard mains luminaires so there is a wide selection in lighting units. The system therefore offers huge flexibility in choice of components as the mainstream luminaires used in standard networks can be employed.

The central or static inverter system has certain advantages and disadvantages.

Advantages

- Little restriction in choice of luminaires as standard mains units are used.
- No problems exist with high or low ambient working temperatures.
- A high lumens output is achieved at extended luminaire spacings in relation to other emergency lighting systems.

Disadvantages

- As with any central battery system variant cabling must be protected although voltage drop is less of a consideration.
- Greater battery power is needed to drive mains lamps in relation to dc output systems.
- Control gear losses can be up to 50%.

The essential operating technique of the central inverter system is as depicted in Figure 5.4

The static inverter cubicle and control gear at which the dc power is converted to ac is extremely robust and although principally designed for use with emergency lighting they can also be found satisfying other electrical loads. This load is subject to a short period break in the supply at the instant of changeover to the inverter from the mains and this should be compensated for when driving equipment other than emergency luminaires.

In the same context as all central battery systems there is a need to use fire protected cable and to give mechanical protection to all of the cabling system. However, by using power factor corrected luminaires in conjunction with a 240 Vac working voltage the level of voltage drops is much reduced. This power factor correction at the luminaire is necessary to minimize the distribution current and any fluorescent circuits with a leading power factor are to be avoided. Equally if low power factor luminaires are used then it is a requirement to bulk correct this at the inverter.

In order to assess the inverter size it is necessary to determine the total output and to include ballast losses together with lamp loads. A further consideration is to verify the types of electronic starters that are in use because although most are acceptable certain versions do cause a direct current to be drawn from the inverter because of their rectification and this presents an overload condition.

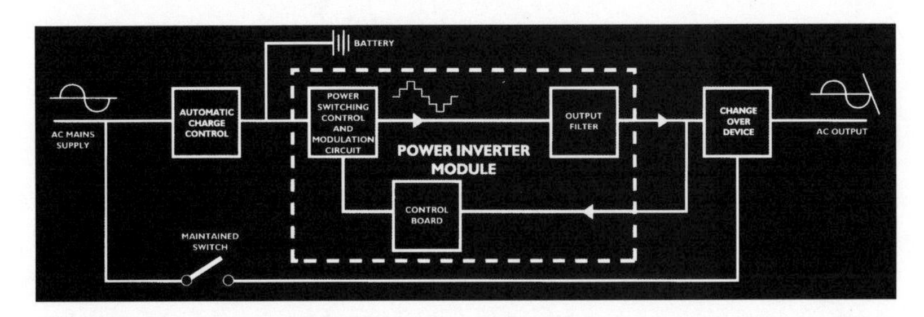

Figure 5.4 *Central inverter system*

In all cases it is necessary to select an inverter that can cater for all of the circuit watts to include the sum of the lighting loads plus any circuit losses.

Inverters will always have a number of options and additional features but they all have a high voltage lockout to interrupt the charger supply to ensure that the battery is protected against overcharging. In addition a low voltage disconnect circuit protects the battery from deep discharge and inhibits the inverter from operating at a level below the design voltage.

We can overview a specification with respect to the details of the instrumentation:

Power range	250 V. 30 kVA
Mains output	240 Vac, 50 Hz single phase +/−10%
Maintained input	Essentially as mains output
Inverter output	240 Vac, 50 Hz single phase +/−6%
Changeover time	2 seconds to achieve full power output
Frequency stability	50 Hz +/−0.5%
Harmonic distortion	Maximum 5% with linear load over entire dc input voltage and power factor ranges
Power factor	0.85 lagging to unity
Overload characteristics	150% load for 1 minute 125% load for 15 minutes
Charger type	Constant voltage and current limited. Temperature compensation control
Battery protection	Low voltage disconnect circuit and high voltage lockout
Contactor	To BS 5424
Fuses	To BS 88
Instrumentation	Mains on indicator Maintained supply on indicator Inverter running indicator Charge ammeter Battery voltmeter Ac output voltmeter Ac output ammeter
Audio visual alarms	Charge failure Mains failure High volts dc Low volts dc

Having accepted that the main advantage of the inverter system is that it can use the large variety of mains lighting units by this it lends itself to be able to function in any general lighting scheme. The need to introduce an extra low voltage scheme is unnecessary and it will be found that the light levels from the inverter powered luminaires are considerably higher than those from traditional extra low voltage emergency lighting systems. The emergency lighting from an inverter system can also be easily integrated alongside the normal lighting particularly in corridors and on staircases to achieve a good aesthetic installation.

The wiring of a central inverter maintained system is illustrated in Figure 5.5.

It will be seen that the load is derived from the normal mains supply via the bypass contactor which is integrated in the static inverter cubicle. This load to the luminaires can be switched both on and off by the use of the maintained remote switch link or by a further switch in parallel at a remote point.

The battery is permanently charged by the charging circuit during normal mains healthy conditions but if this supply fails the contactor connects its output to the lumnaire loads. In the event that the rated duration of the battery is exceeded the low voltage disconnect circuit functions. Once the mains is restored the contactor connects the load to the bypass circuit and the battery begins to recharge. Effectively the inverter power circuit is only used when the mains fails so the inverter is classed as operating in a passive standby mode.

In systems that use an inverter together with luminaires that are combined for both mains and emergency use it is possible to have individual control at a local point. This is carried out by means of local switches and a supply to the lamp by a changeover device so that the

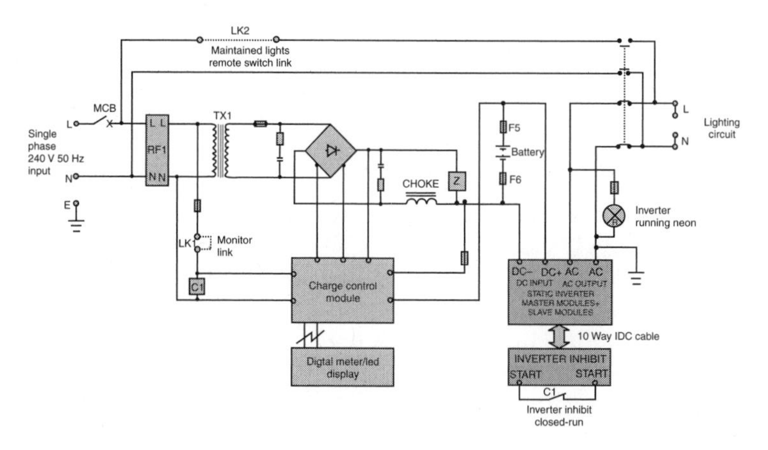

Figure 5.5 *Central inverter maintained system*

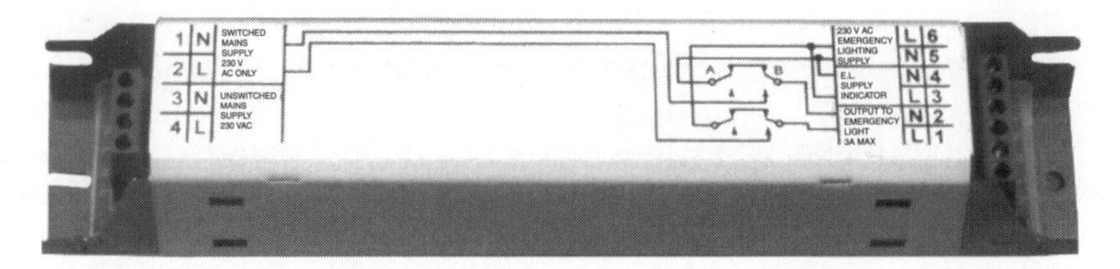

Figure 5.6 *Changeover device*

lamp is wired to the local switched mains supply during normal service but from the static inverter under mains fail conditions. These devices are installed within the luminaire and are engineered to switch 480 volts as the inverter could be on a different phase to the changeover device. If the device cannot be easily fitted within the luminaire it can be housed in a remote location. The wiring is as shown in Figure 5.6.

A range of slave luminaires exists specifically for central inverter systems and these are power factor corrected so that the size of the static inverter can be controlled. These luminaires also have electronic starters since EN 60598-2-22 prohibits the use of glow starters in units when the starter is in circuit during emergency operation. These luminaires will be catalogued typically as in the following example with respect to a typical 8 W slimline fitting.

Model	*Input voltage*	*Lamp type*	*Ac power input*	*Power factor*	*Light output*	*Wt*	*Size*	*Temperature range*
SL8	230/240	8 W	16 VA	0.9	420 lumens	1.25 kg	410 × 150 × 95 mm	0–25°C

The range of fittings will extend through edge lights and exit luminaires to recessed lamps, circular fittings, slimline units, square decorative lamps and weatherproof versions with a level of vandal resistance. Many manufacturers offer high frequency mains luminaires for use with inverter systems which have an instant start high efficiency ballast to meet BS EN 60598-2-22.

Electronic starters are currently fitted as standard in most fluorescent luminaire ranges and are offered as an option in most others. They contain no moving parts or contacts to wear out and remove the need to replace starters for the life of the light unit. In addition electronic start greatly enhances tube life to the extent that blackening and life reduction caused by switching becomes negligible and the full rated burn life of the tube can be achieved. Luminaires fitted with

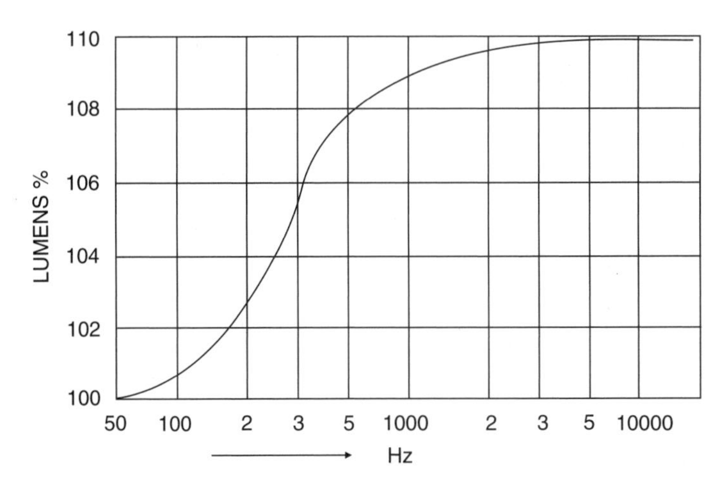

Figure 5.7 *Variation of lamp efficacy with operating frequency*

electronic start will also operate at lower temperatures that can be as low as –35°C. Electronic starters also improve the time taken to ignite the lamp and by generating one positive pulse the lamp can be ignited in less than 3 seconds from switching.

With regard to high frequency luminaires it can be established that a characteristic of fluorescent lamps is that their efficacy increases with frequency. This efficacy or lamp circuit luminous efficacy is expressed in lumens/circuit watt and is the ratio of the luminous flux emitted by a lamp to the power consumed by the lamp with the power consumed by the control gear taken into account. Increasing the frequency of operation from that of the normal mains supply of 50/60 Hz to typically 30 Hz improves lamp efficacy by approximately 10%. High frequency ballasts are designed to have lower power losses than most conventional wire wound ballasts giving a further increase in current efficacy. Due to the reduced power consumption of the high frequency system the temperature within high frequency luminaires will be lower than within mains frequency luminaires. Compared to mains frequency ballasts power loss in high frequency ballasts can be reduced by up to 65%. Figure 5.7 shows the actual variation of lamp efficacy with operating frequency.

At the stage of establishing all of the luminaires to be used with the central inverter and incorporated in the system it is necessary to list the characteristics of the lamps and to consider in detail the fluorescent loads as these operate at a power factor less than unity. The fittings and VA ratings should then be charted typically as shown in Table 5.1.

Table 5.1 *Charting light fittings VA total*

		VA rating		
No.	*Type*	*without pf corr.*	*with pf corr.*	*Total circuit (watts)*
25	58 W single	2050	2050	1775
40	28 W 2D	3080	1694	1440
12	16 W 2D	576	296	252
5	40 W incandescent	200	200	200
14	PL-11	538	247	210

It can be noted that:

- VA rating lamps without power factor correction = 6.444 kVA
- VA rating lamps with power factor correction = 4.487 kVA
- Total circuit W = 3.877 kW

From this it may be established that because of the poor power factor of uncorrected 2D and PL-11 luminaires as specified an inverter rating of 6.4 kVA is required to support a load of 3.877 kW.

If the power factor was actually corrected to 0.85 lagging the inverter size could be reduced to 4.487 kVA although the total circuit watts would remain at 3.877 kW.

In general when selecting an inverter system for a building for simplicity it is normal to use a single large system. However, for large sites it is often better to use a number of smaller units and locate these at different but convenient points. In these cases the failure of an individual unit cannot compromise the function of the entire network. There are a number of points that should be checked and logged before establishing the system to be used:

- Consider the full building size, number of floors and the length of cable runs particularly to the furthest luminaire as this has a major influence on the system voltage. The actual influence of the system voltage due to the length of cable runs must not be overlooked.
- Calculate the load of the system. Make compensations for variations in battery performance and for any future expansion of the system.
- Log the power factors of the luminaires. Typically as shown in Table 5.1.
- Determine the mode of operation.
  – *Non-maintained*. When the system ac fails the battery is automatically connected to the lighting load to give a dc supply.
  – *Maintained*. Provides a permanent supply. Under mains healthy conditions the ac supply is provided by a stepdown isolating

transformer. In the event of a mains failure the luminaire load is connected to the battery.
– *Floating system*. The dc supply is permanently connected to the load. The floating emergency supply is connected when a mains failure is sensed. The charger is rated in the system to control the battery and load current independently.
– *Static inverter ac central system*. The ac load is normally derived from the mains through a contactor. If the mains fails dc battery power is converted to ac at mains frequency by a static inverter to supply the load. Three phase supplies are available for heavier loads.

- Determine the battery type to be used and the room in which the cubicle is to be sited. This must always be in an area of low fire risk.
- Make a log of all of the options that may be required in the full network.
- Plan the wiring runs and the output distribution for all of the circuits to include the fusing.

As a conclusion we may say that the central inverter system has a great deal to offer because of its ability to use standard mains luminaires and this enables high illumination levels to be achieved. In addition for high security risk areas the security lighting networks can also be linked to the inverter if spare capacity is available. This ensures that the security lighting may still function in the event of a mains disconnection. However, it remains of vital importance to calculate the loads alongside the control gear losses and to ensure that the most efficient luminaires are used to minimize the total distribution current.

6 Wiring systems

The cabling associated with central battery systems is much more involved than that of the self-contained system. In the case of the central battery system, because the source to power the luminaires is always at a remote point the wiring has to be protected against fire and mechanical damage as a result of impact or working activities that could damage the cables. The system therefore relies on the integrity of the cables to carry the power to the lighting units under all working and emergency conditions. The self-contained system uses battery backs that are integrated within the luminaires so the cabling is therefore only the same as that used for the normal lighting since if it becomes destroyed by fire the emergency supply will operate. The wiring for the self-contained luminaire is only governed by the normal standards needed for protecting and running in the cables that are applied in the electrical engineering sector for the particular premises and building type. However, the wiring needed for central battery systems including central inverter systems is more onerous and is considered in its own right in the following section.

The requirements of the IEE Wiring Regulations and codes of practice have always been compiled with a condition that circuits are to be designed against available data and this must be available for reference. However, before it is possible to calculate cable size it is necessary to determine the cable type to be used and the method of installation and the environment in which it is to perform. In practice there are a number of stages in the procedure for the calculating of the cable performance but in the emergency lighting sector we are not so concerned with high levels of power such as those found in many other electrical engineering applications. Nevertheless these time-honoured stages can be used to achieve a judgement in relation to emergency lighting systems:

- Determine the design current. This can be obtained from the manufacturers' data or from the rating of the equipment on the installation sheets.
- Select the rating of the protection. This can also be taken from the manufacturer or selected from their charts. The fuse or MCB is of vital importance in the protection of the cables.

- Recognize that the cable current carrying capacity is influenced by the correction factor as this derates the capacity or must conversely increase the cable size. For conditions such as high ambient temperature, cables grouped closely together, uncleared overcurrents and contact with thermal insulation, the cable when carrying its full load current will become warm. It is important to note that in the larger premises cables are often run in close proximity and may also be fully loaded electrically at the same point in time. It may be noted that in general cable ratings are based on an ambient temperature of 30°C so above this limit a correction is needed.
- Take notes of thermal insulation because this may be in contact with cables and if it totally surrounds a cable a derating value has to be applied.
- Take particular note of voltage drop. The resistance of a conductor increases as the length increases and/or the cross-sectional area decreases. This results in a drop in voltage: the load at the end of a relatively thin cable cannot have the full supply voltage available. This voltage drop is not to be so excessive that the equipment can not function correctly.

6.1 Protection against fire and mechanical damage

Cables must always be routed through areas that are of low fire risk and be segregated from the wiring of other systems.

The cables and wiring systems that may be employed are:

- Cables that have an inherent high resistance to attack by fire.
  - Mineral insulated copper sheathed cable (MICC) in accordance with BS 6207: Part 1.
  - Cable in accordance with BS 6387. This cabling should be at least category B.
- Wiring systems that require additional fire protection.
  - PVC insulated cables in accordance with BS 6004 in rigid conduit.
  - PVC insulated cables in accordance with BS 6004 in steel conduit.
  - PVC insulated and sheathed steel wire armoured cable in accordance with BS 6346 or BS 5467.

In all cases the systems are to be cabled to satisfy the IEE Wiring Regulations and BS 5266. It is to be appreciated that additional fire protection is always afforded if the cables are buried in the structure of the building. MICC cable is worthy of special mention because of its unique role.

Mineral insulated copper sheathed cable (MICC)

This cable consists of conductors of solid copper insulated with a highly compressed covering of magnesium oxide (MgO) powder and sheathed with a seamless malleable copper tube. It is extremely robust yet pliable enough to be bent easily. Its main advantages can be summarized:

- Strong and robust, it does not require further mechanical protection unless it is installed in an area where it is liable to be impacted or damaged.
- It can withstand very high temperatures and is non-flammable.
- The conductors held within the cable are not disturbed in relation to each other by bending or twisting the cable.
- It has good resistance to oil, water and condensation when the ends of the cable are sealed correctly.
- Life expectancy of the cable is good.
- In relation to the size of the cable the current carrying capacity is high.
- The copper sheath gives good earth continuity.

Copper MICC cable can be run along the surface of walls and other structures and can be fixed in position by the use of copper clips and saddles. Ferrous fixings should not be used because they can create a chemical action under damp conditions. These cables can also be laid under floors or they may be buried in concrete, plaster or in the ground, but if exposed to the weather or to a risk of corrosion or if laid underground or in a concrete duct they should have an overall PVC sheath. This sheath gives protection against mechanical damage but if the cables are run at a low level on the surface it is advisable to enclose them in conduit or channelling.

The bending of this cable type can be done by hand or by means of purpose designed tools. The bending radius should be limited to six times the diameter of the cable as this will enable the cable to be straightened at a later point if so desired without any excess distortion. In order to avoid damaging the cable sheath the tools used to form the cable are best faced with hard leather so that the cable sheath is not damaged. In addition a wooden block and hammer can be used to dress the cable to improve the neatness of its final appearance. The cable ends must always be sealed during installation as moisture can penetrate the assembly slowly and have a detrimental affect on the compressed covering.

In order to effectively seal the cable ends attention must be made to the making-up of the termination. This is because magnesium oxide is absorbent.

The assembly comprises two subassemblies, namely the seal and the gland with each performing a different function. The seal is intended to prevent moisture entering the cable and the gland is used to anchor the cable. The seal is made up of a brass pot with a disc to close the assembly entrance and the sleeves are used to insulate the conductor tails. A compound is used to fill the actual pot which is screwed into position. The finished assembly is shown in Figure 6.1(a). The most used gland type is the compression gland and this is used to hold the assembled cable at a joint box or at the termination point. This is depicted in Figure 6.1(b). A range of glands and lock nuts is available for the different entrances that may apply for the cable termination points.

One of the main advantages of MICC cable when used in central battery systems is the fact that it has higher current ratings than PVC or rubber insulated cables of the same conductor size. Therefore for a given current rating an MICC cable will tend to have a smaller conductor size and overall outside diameter. The reason for this is attributed to the high thermal conductivity of the magnesium oxide insulation that enables the heat generated in passing current to be

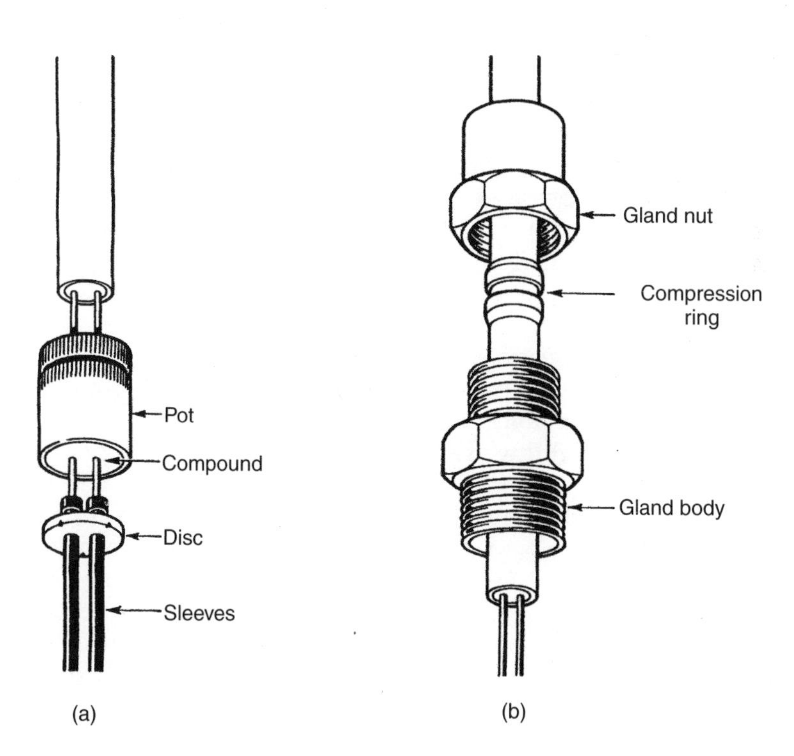

Figure 6.1 *(a) MICC cable seal. (b) MICC cable gland*

quickly discharged from the cable. The characteristics of this cable are shown in Table 6.1 which refers to a pair of conductors. The values in brackets relate to manufacturers' standard figures for general PVC insulated copper conductor wiring systems and can be used to illustrate the benefits of MICC cable.

In theoretical terms a voltage drop in any cable run should be determined before specifying any system voltage. This is easily calculated and can be measured in a practical way once the system has been installed.

Total volts drop = load current max. × volt drop per amp per metre × cable length in metres

The volt drop per amp per metre is obtained from the manufacturers' data for a given cable type.

If we consider a 50 V system with a power consumption of 1650 W then W = VA or 1650 = 50 × A. From this the system will have a maximum load current (A) of 33 amps. If it is intended to use 6 mm cable over a distance of 45 m then by reference to Table 6.1 the calculation would be:

$$\text{Total volts drop} = 33 \times 7 \times 45$$
$$= 10395\text{ mV or } 10.4\text{ V}$$

A voltage drop of 10.4 volts may be unacceptable in a 50 volt system so it is possible to consider a different system voltage. As an example for a 110 volt system the maximum load current would be 1650/110 = 15 A. The calculation then would be:

$$\text{Total volts drop} = 15 \times 7 \times 45$$
$$= 4725\text{ mV or } 4.725\text{ V}$$

This is a more acceptable value.

In emergency lighting there is a need when selecting cable sizes to pay special attention to the limitations that are imposed by both voltage drop and the physical strength of the cable. The voltage drop in any cable forming a connection to a slave luminaire from a central battery cubicle is not to exceed 10% of the system nominal voltage at the maximum rated current.

We accept that by using the mathematical term $V = IR$ the current can be predicted in a circuit when we know the terminal voltage and the combined circuit resistances. Calculations can always be verified in the field to prove that there is no fault such as leakage to earth. In practice, when faced with voltage drop considerations, most installers

Table 6.1 *Copper MICC cable characteristics*

Nominal cross-sectional area of conductor (copper) (mm²)	*1 × 2 core single cable single phase ac or dc*	
	Current rating (A)	*Volt drop per ampere/metre (mV)*
1.0	19.5 (14)	42
1.5	25 (17)	28
2.5	34 (24)	17
4	45 (32)	10 (11)
6	57 (41)	7 (7.1)
10	77 (55)	4.2

in the electrical sector double up the supply cores to reduce the resistance of the supply conductors. We know that the greater the distance, or the smaller the cable, the greater is the resistance, and as the voltage at the terminals of the supply is fixed, a voltage must be lost over the cable run. In addition damage to cables and poor joints themselves lead to a rise in the resistance of the supply conductors and a reduction at the head voltage or the furthest point in the circuit. This head voltage is important and is effectively the terminal voltage less the sum of the voltage drops in the circuit. It can be both calculated and measured and the results compared. This is to be appreciated since there is an operating range for every luminaire and the effects of going under the working level inhibit correct operation.

Mineral insulated wiring cables can be supplied with an outer plastic covering of a low smoke zero halogen material or simply with a bare copper sheath. They will not give off any smoke, acid gas or toxic fumes. Equally they do not add to the spread of fire and will not provide any fuel for it and can operate indefinitely at temperatures up to 250°C. In addition the densely compacted magnesium oxide insulation will prevent the transmission of explosive gases between equipment without any further sealing. For all of these reasons this cable type provides a safe, durable, high integrity installation and is actively promoted for safety critical circuits required to operate under fire conditions to include emergency lighting supply circuits.

Standard fire performance cables

MICC cables have a long history of use in circuits intended to operate under fire conditions but as technology has advanced and buildings have become increasingly complex the consequences of fire have become greater and this has resulted in optional fire performance cables becoming available. The need for systems requiring cable to operate under fire conditions has never been greater and there has also been a tendency for new generation cables to be more easily terminated than MICC to reduce installation times.

There is now a range of manufacturers producing a comprehensive standard range of fire performance cables which are designed to continue to operate and emit minimum quantities of smoke and toxic fumes in the hazardous conditions of a fire. The terminating of these cables is both quick and simple and requires no special tools for bending or removing of the sheathing. Each core is colour coded to aid identification and the composite sheath provides excellent mechanical properties. These can often be surface or subsurface mounted without any additional protection from impact or mechanical damage. However, if conditions dictate and the cables are liable to be damaged or are in accessible areas they can be purchased in an armoured cable version with wire armouring under the outer sheath to give extra high resistance to abrasion and impact.

Two of the most prominent standard yet robust fire resistant cables are FP200 (a trademark of Pirelli General Cable Ltd) and Firetec (a trademark of AEI Cables Ltd).

FP200

This cable features conductors of plain annealed copper in a silicone rubber insulation with an earth conductor of tinned copper. It has an aluminium laminate and LSOH composite sheath that is available coloured red or white and provides excellent mechanical properties. These cables are small but strong and can be used in high humidity and moisture environments and meet the requirements of BS 4066 and relevant IEC standards for flame retardant cables. This cable can operate at 70°C although the insulated cores will function at 150°C. The assembly needs no special tools for bending or removal of the sheathing and each core is colour coded to aid identification. A termination ferrule is used at the final connection points.

The colour coding is:

2 core Red black
3 core Red blue yellow
4 core Red blue yellow black

Table 6.2 illustrates the size of the cable by cross-referencing the cable overall diameter against the number of cores related to the solid conductors.

Table 6.2 *FP cable overall diameter at different cross-sectional areas*

	Cable overall diameter (mm)		
No. of cores	*1 mm^2*	*1.5 mm^2*	*2.5 mm^2*
2 + earth	7.9	9.0	10.5
3 + earth	8.2	9.5	11.4
4 + earth	9.0	10.2	12.4

Table 6.3 *FP200 current rating*

Cross-sectional area	*Conductor Size*	*Current rating (A)*	
		Note 1	*Note 2*
1 mm^2	1/1.13 mm	13	15
1.5 mm^2	1/1.38 mm	16.5	19.5
2.5 mm^2	1/1.78 mm	23	27

Note 1. Cable enclosed in a conduit or trunking (conductor operating temperature 70°C).
Note 2. A single cable in free air or embedded direct in plaster (conductor operating temperature 70°C).

Table 6.3 illustrates the current rating.

Firetec

This is an alternative progressive cable type for use in industries and applications that will enable the safe and orderly evacuation of personnel during emergency conditions. It is available as a multicore in 2, 3, 4, 7, 12 and 19 core variants in popular conductor sizes with the outer sheath in either red or white as standard.

Firetec meets the highest categories of BS 6387 – C, W and Z – and is LPCB certificated. It is easy to install and needs no special tools and is of low smoke, zero halogen construction. Firetec is designed and manufactured under an ISO 9001 quality management system. The conductor insulation gives low capacitance values and high mechanical strength. The cable has stranded copper conductors for increased flexibility in preference to a single conductor.

The cable assembly comprises stranded plain annealed copper conductors surrounded by a mica/glass tape which is held within a tube of cross-linked insulation. This is held in a wrapping of aluminium/polyester tape and the entire assembly is surrounded by a tough outer sheath. A circuit protective/drain wire is also provided.

The manufacturer claims that this cable type is ideally suited for use on emergency lighting circuits and satisfies the requirements of the relevant code of practice BS 5266 Part 1. It also claims to meet the requirements of numerous other standards for fire resistance, flame retardancy, smoke density and acid gas evolution and the maintenance of circuit integrity when subject to a 750°C flame to IEC 331. It also satisfies BS 6387 to categories C, W and Z.

Category C — Resistance to fire alone (950°C for 3 hours).
Category W — Resistance to fire with water spray (650°C for 15 minutes with fire alone followed by fire and water spray for 15 minutes).
Category Z — Resistance to fire with mechanical shock (950°C with applied mechanical shock for 15 minutes).

To install Firetec requires only a special purpose gland to terminate the cable at the equipment points. The practices otherwise are general with respect to the IEE Wiring Regulations and relevant codes of practice. As with terminating FP200 the installation is quicker and easier than with MICC cable. The minimum bending radii should not be less than six times the overall cable diameter and the fixing is to be in line with the manufacturers' recommendations so that as a general guide the distance between fixings is 300 mm. One-way clips are used to support the cable when surface wiring and these are of copper with a plastic coating to give them a fire performance equal to the cable. Table 6.4 gives an indication of the data and performance of such a cable up to a 4 core construction.

It is noted that the table provides a value for volt drop (mV/A/m). We have established that this is mainly applicable to low voltage cables and can be calculated using the formula for dc and single phase ac two wire systems:

$$V = (2 \times I \times L \times R)/1000$$

where I = current carried per conductor in amps
L = length of cable run in metres (one conductor only)
R = resistance in ohms as given in the manufacturers' appropriate tables

For convenience the voltage drop factor as given in Table 6.4 requires only multiplying by length and current to give the voltage drop in volts.

Armoured cables

As an extension to using Firetec fire performance cable an armoured version is available. This is not unlike the former type but also has a

Table 6.4 *Performance data fire resistant cable*

Conductors		Current rating (amps)	Volt drop (mV/A/m)	Max. dc conductor resistance at 20°C (ohms/km)	Nominal overall (mm)
No.	Cross-sectional area Cores (mm^2)				
2	1	15	44	18.1	8.2
2	1.5	19.5	29	12.1	9.2
2	2.5	27	18	7.41	10.1
2	4	36	11	4.61	11.4
3	1	13.5	38	18.1	8.6
3	1.5	17.5	25	12.1	9.9
3	2.5	24	15	7.41	11.0
4	1	13.5	38	18.1	9.6
4	1.5	17.5	25	12.1	10.9
4	2.5	24	15	7.41	12.0

steel wire armouring within the outer sheath to provide extremely high levels of resistance to abrasion and impact. The installation methods and procedures for this cable are the same as those that would be adopted for standard steel wire armoured cable. Standard cable glands to BS 6121 are used although if onerous conditions are expected to exist then this is to be taken into account when selecting the gland to be used. Steel wire armoured cable is used in applications where a high degree of mechanical protection is needed. It may be fitted directly onto a wall, placed in ducts, or directly buried into suitably prepared ground. They tend to be approved to BS 6346 and will have different current ratings depending on whether they are buried directly in the ground, run in a duct or clipped in free air. There are a number of different materials that are available for forming the external sheath ranging from standard PVC through to a range of flame retardant materials.

Although we have come to understand armoured cables to BS 5467 and BS 6346 it is also to be known that new generation ranges of cables are available that have non-armoured high impact sheaths of XLPE insulation to BS 5467 or PVC insulation to BS 6346. XLPE is an insulation of cross-linked polyethylene applied by extrusion to form a compact homogeneous layer to provide low capacitance values and high mechanical strength. All of these different cable types come in a range of conductor sizes and number of cores and once again need no special tools to terminate the cable ends.

Although many cables are installed in areas unaffected by impact or abuse there is still an extensive use of conduits to afford additional protection and to improve the aesthetic values of an installation.

Rigid conduit

The use of polyvinyl chloride (PVC) as a rigid conduit in the electrical engineering sector is extensive. It is essentially a tube or pipe intended to hold and protect the conductors.

The most commonly used type of rigid conduit is 'unplasticized' PVC conforming to BS 4607. This has plain bored ends and is generally self-coloured in black or white.

When installed it offers a degree of protection against mechanical damage. The normal method of joining and applying fittings is by the use of push fitting. This push fit entry technique ensures a tight, reliable fit and when used in conjunction with PVC adhesive a strong permanent joint is achieved. For additional protection in damp environments solid rubber gaskets can be added to the assembly. The threading of PVC conduit is not normally attempted because it weakens the conduit wall.

There is a huge range of fittings and accessories available for use with rigid conduit systems. These include couplings and reducers through to bushes, T pieces, bends and elbows with a vast variety of junction boxes for use in line with the conduit runs. These can also be used in conjunction with inspection bends that have detachable covers. Such fittings and accessories make installations as easy as possible and enable branch conduits to run off from each other. The inspection and drawing of cables can also be accomplished at a multitude of points.

The coupling of rigid conduit is easily carried out by the application of plain bore push fit components to the conduit tubes. Solvent adhesive can then be applied to improve strength and watertightness if need be. Conduit fittings are also available with special sockets and locking rings to hold conduits in place at such points as end boxes. Expansion couplers are used where temperature variations of 14°C are expected as these allow the conduit to move in and out as it expands and then contracts. Cold bending of PVC can be achieved on tubes not greater than 25 mm diameter and bending machines are available for this purpose. As an alternative bending springs can be used. It is also possible to apply heat to the PVC conduit and then insert a bending core for slight bends to be achieved. It is normal to overbend slightly since PVC tends to straighten out after bending or cooling. With all PVC conduits accessory boxes can be used and connection is made by the use of PVC adaptors which may be clip-in plain bore, or screwed at one end and plain bore entry at the other end. The screwed types may be male threaded for use with locking rings or female threaded for use with male bushes.

To summarize the advantages of rigid PVC conduit:

- High tensile strength is offered.
- The electrical resistance is high. The electrical breakdown voltage is in the order of 12–20 kV/mm.
- If warmed slightly it can be bent.
- Installation is not labour intensive so the fitting procedure is inexpensive.
- It has a good degree of protection to the weather.
- When used at its specified working temperature ranges it will not crack.
- It is of low flammability and is self-extinguishing when the flame source is removed.
- It has high resistance to corrosion by water, acids, alkalis and oxidizing agents. These materials are also unaffected by the chemical compounds in concrete and plaster.
- It does not cause any significant degree of condensation.
- It is dimensionally stable.
- It does not deteriorate significantly with age or external exposure.

The effect of high and low temperatures must be considered when using PVC. Rigid PVC conduit and the fittings intended for use with it are not suitable when the ambient temperature can fall below –5°C or where the normal working ambient is greater than 70°C. In those cases where a marginally elevated or lowered temperature is only present for a short period of time it is possible that this will have no detrimental effect on the conduit but the manufacturer should be consulted. In low temperatures the PVC becomes harder and less ductile and is then more susceptible to impact. PVC expands some 12 mm for a 4 m length against a temperature rise of 45°C and for this reason expansion couplers should be used when rigid conduit is installed in straight runs for lengths in excess of 5.5 m. An exception to this is when bends are employed as these compensate for the expansion. The saddles that are used as supports for the conduit can be set to permit a measure of lateral movement.

The fitting of rigid conduit is relatively straightforward and in the installation process saddles and clips must be fitted to allow longitudinal movement to cater for temperature changes in the environment. The spacing of supports is shown in Table 6.5. If the ambient temperature is high or if the working area is subject to rapid changes in temperature the fixing centres are to be suitably reduced. PVC causes no significant

Table 6.5 *Spacing supports for rigid conduits*

Normal conduit size (mm)	*Maximum distance between supports*			
	Rigid		*Flexible*	
	Horizontal (m)	*Vertical (m)*	*Horizontal (m)*	*Vertical (m)*
< 16	0.75	1.0	0.3	0.5
> 16 but < 25	1.5	1.75	0.4	0.6
> 25 but < 40	1.75	2.0	0.6	0.8
> 40	2.0	2.0	0.8	1.0

internal condensation so drain holes need not be applied. Fixings should be 150 mm from bends and good aesthetics can be achieved by keeping supports in long runs equidistant.

Included in this table are the fixing distances for flexible PVC conduit. This is available in long coiled lengths and would only be used in emergency lighting systems to give protection to cables in sunk applications where floors are awaiting screeds. It is sometimes used before burying in shallow plaster.

An alternative to PVC rigid conduit is steel conduit.

Steel conduit

This type of tubing provides excellent resistance to fire and because it can be both threaded and bent it makes up an extremely versatile wiring system. In general it is galvanized or covered with black enamel paint to suit the premises in which it is to be installed. A variety of fittings are available so it is possible to incorporate it into a complete installation system.

One of the main considerations of using steel conduit is that the correct tools, accessories and methods of preparing the conduit must be accounted for as it uses a more time-consuming assembly technique than that of the making up of rigid conduit systems. This is because of the need to apply screw threads to the tubes that are then screwed to additional tubes using couplers or are screwed to the many accessories available. These accessories perform a similar duty to those that are used in PVC conduit installations.

In order to make up any steel conduit system there is a need to have a conduit-bending machine and pipe vice together with stocks and dies.

In order to thread a length of steel conduit a pipe vice is used to hold the tube which is then cut to the desired length, slightly chamfered and then threaded using the stocks and dies. Any particles of steel debris should be removed from the inside of the tube and the thread should not be painted before it is screwed to its recipient. This is because the conduit is normally used as a circuit protective conductor. If the conduit is to be bent the radius of any bend is not to be less than 2.5 times the outside diameter of the conduit. It is important to employ a number of inspection type fittings in the assembly for the removal or addition of cables.

If non-inspection elbows and tees are used they are only to be installed at the end of a conduit immediately behind a luminaire or outlet box of the inspection type. As an alternative a solid elbow can be installed at not more than 500 mm from a readily accessible outlet box in a conduit not exceeding 10 m between two outlet points providing that all other bends do not form more than 90°. It is also important with steel conduit systems to ensure that cables cannot be damaged when they are pulled into position so bushes must be used in boxes and at entries to accessories and luminaires etc. A further consideration is that drainage holes must be provided at the lowest points of an installation to enable moisture due to condensation to drain away.

The main advantages of steel conduit are apparent:

- Excellent protection afforded to the cables against mechanical damage and impact.
- Excellent resistance to flames and fire.
- Provides earth continuity.
- Is not easily affected by extremes of temperature.

There are a number of methods used to fix metal conduit in position. If the conduit is to be covered by plaster crampets are used to secure it in place before this procedure is carried out. Crampets are similar to a standard cable clip but of greater proportions so that they can hold the conduit in position in preparation for the plastering. If there is a need to surface fix the conduit saddles may be used. These hold the conduit in position against the fixing surface and can have spacer bars added to them to hold the conduit slightly off the surface. Distance saddles can be employed if there is a requirement to hold the conduit sufficiently from the fixing surface to prevent dust from gathering between the conduit and the fixing surface.

Table 6.6 shows the spacing supports for rigid metal conduit.

Steel conduit in addition to being suitable for fixing to finished walls and building structures can also be fixed to unplastered walls or

Table 6.6 *Spacing supports for rigid metal conduit*

Nominal conduit size (mm)	Maximum distance between supports	
	Horizontal (m)	Vertical (m)
< 16	0.75	1.0
> 16 but < 25	1.75	2.0
>25 but < 40	2.0	2.25
> 40	2.25	2.5

chased into the building structure before being encased in plaster. In the latter case it is necessary to allow the plaster to dry before drawing any cables into the conduit to prevent the ingress of any moisture into the tubes from the wet plaster.

Steel conduit is available in standard lengths of 3.75 m in diameters of 20 or 25 mm conforming to BS 4568. For aggressive environments high grade 316 stainless steel conduit is available; this type of conduit is able to cope with chloride-laden conditions and is ideal for applications in the food, brewing, dairy, pharmaceutical and chemical industries. Stainless steel also has an aesthetically pleasing finish and can be used for architectural reasons in emergency lighting systems where surface wiring must be applied and appearance is important. As with PVC conduit waterproof installations can be achieved but in this case by the adoption of screwed joints and watertight fittings. It must be remembered, however, that all screwed fittings are not of necessity watertight so when this property is required it must be clearly specified.

Choice and erection of conduit

Steel conduit has been in existence for many years and is considered an essentially sound system. It offers complete protection of the cables and the screwed joint is secure so that the conduit is electrically and mechanically continuous. The recommended installation practice is for the conduit to be erected in position prior to the cables being installed and this then complies with IEE (Institute of Electrical Engineers) regulation 522-08-02. After considering the plans of the installation the erecting is achieved by first taking the main runs from the location of junction boxes and draw-in boxes to where the branch circuits radiate. The branch run will carry the cables that supply several lighting points together. The same method is then chosen for the adjacent lighting points. Draw-in boxes are included in the runs as necessary to allow the drawing in of the cables. When erecting the conduit it is important to avoid overcrowding of cables in the tubes. Excessive strain applied when drawing any cable into a tube that is overcrowded may cause some of the strands of multi-stranded cable to fracture and thus lower the current carrying capacity of the cables; the cable may become stretched or even detached; and the insulation can be damaged. In a well-planned installation it should be possible to draw from the conduit any defective cable without disturbing any of the other cables in the same tube.

Conduit can be used to give additional protection to any cable type and to improve the aesthetics of an installation. For this reason it can be used to enclose fire performance cables when they pass through areas that could present a risk of damage to the cables or impacts being applied to them. In every emergency lighting installation it is necessary to plan the cable runs so that the cables are not in accessible areas and are not at points in which they could be damaged or impacted when additional engineering works or maintenance are being carried out. This even applies in ceilings and voids which are most likely to be used on future occasions to run in extra services such as pipework or additional electrical services and cables. The emergency lighting cabling for central battery systems is to be segregated and not run with other cables supplying the general business services and therefore is not to be put in the same flange trays, trunking or go through the same holes through walls as the other cabling systems.

We have taken note of BS 6004 cables. This refers to single core, twin or multicore PVC insulated cables, non-armoured with or without a sheath. We have also established that the insulation which immediately surrounds a cable conductor is designed to withstand the cable's working voltage in order to prevent danger. The additional sheathing and/or armouring is added to protect these insulated conductors from mechanical damage.

In accepting that some environments are more hazardous than others, the cables must be chosen to suit that specific environment and its fire performance for safety critical functions. PVC insulated and sheathed cables are essentially suitable for all types of domestic and commercial wiring installations where there is little risk of mechanical damage, extremes of temperature or corrosion. It is not expensive, easy to handle and terminate, is easily understood and needs no specialized tools to work with. On the basis of this it is used for self-contained emergency lighting circuits but does not feature adequate protection against fire or impact for central battery systems unless totally enclosed in conduit throughout its entire length. These tend to be called surface wiring cables and are used extensively in both the domestic and commercial environment plus industrial areas. They have a PVC insulation and PVC external sheath. Conduit cables are not unlike these but are single conductors with a PVC insulation and only intended for use in commercial electrical installations where the cables are to be protected by conduit.

7 Batteries and data

In emergency lighting systems the importance of the batteries cannot be underestimated as these provide the only source of energy in the event of a power cut. We are able to define the battery as one or more cells that are interconnected to form the standby secondary power supply as part of the emergency lighting system. The emf of the battery is generated by the effects of dissimilar metals being immersed in an acid or salt solution known as an electrolyte which is then used to supply the load. However, the chemical qualities and efficiency of the electrolytes deteriorate with use and have to be renewed at a given point in time. Nevertheless the materials used in the production of batteries are becoming more refined as battery technologies advance and allied to this manufacturers recognize a need to engineer equipment to work more efficiently at lower power levels.

7.1 Self-contained systems

In general these networks use either high temperature sealed nickel–cadmium (NiCad) or maintenance-free sealed lead–acid batteries (SLA).

High temperature sealed nickel–cadmium batteries are used extensively in the vast majority of luminaires and illuminated exit signs whereas the sealed lead–acid battery is normally only found in high intensity tungsten halogen twin floodlights. We may recall that the latter luminaire is used primarily in large high risk areas which are too high for self-contained fluorescent units.

The product standard for emergency lighting is BS EN 60598-2-22 and is invoked in BS EN 1838. When considering battery life these standards call for self-contained luminaires to have a battery designed for a minimum life of 4 years with the luminaire operating normally and in the cases of a maintained lamp at 1.06 × rated supply voltage. Measurements are to be made to confirm that battery temperatures are less than 50°C and module temperatures are less than 50°C or as stated by the manufacturer of the goods. In practice modules tend to be rated at a maximum case temperature of 70°C and batteries at up to 55°C for higher rated cells.

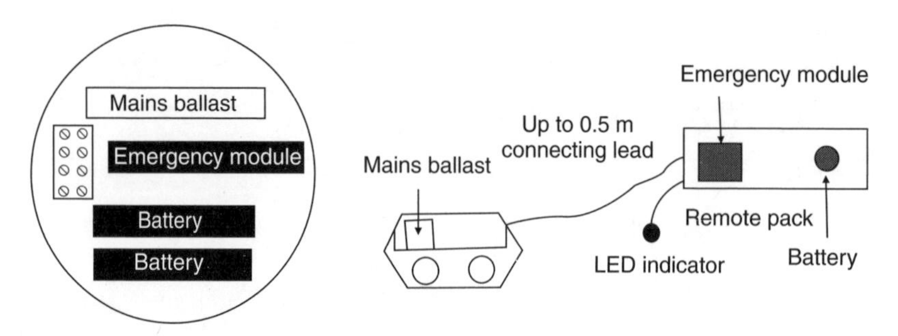

Figure 7.1 *Converted self-contained luminaire*

Luminaires must be carefully designed to ensure that the temperature and electrical stress is within the limits that are specified by the cell manufacturer in order to achieve the required 4 year design life. ICEL registered luminaires must be verified for operation within these limits and also have regular battery audits to verify the life expectancy. The evidence of a 4 year design life in BS EN 60598-2-22 invokes an initial test plus an accelerated life test together with regular ongoing audit tests.

Although there is a vast array of self-contained fittings it may be the case that an existing standard light may be needed to operate as an emergency light. In those cases where it is required to convert a general light or special luminaire for emergency use, this operation can be performed by using a conversion kit. This kit comprises an emergency ballast or module together with the rechargeable battery. In the event that the luminaire has limited space or if the temperature of the luminaire could create a problem for the battery a remote pack can be employed and be mounted within some 0.5 m of the luminaire – see Figure 7.1.

Nickel–cadmium cells

The nickel–cadmium cell has a positive plate of nickel hydroxide and a negative plate of cadmium mixed with a small amount of iron. The electrolyte is potassium hydroxide. The active chemicals in the plates are enclosed in thin nickel–steel grids insulated from each other by ebonite rods and the entire assembly is enclosed in a welded steel container. These cells are readily available in a number of sizes and are easily replaced in self-contained luminaires. They have certain advantages and disadvantages in use.

Advantages

- Totally enclosed so are effectively gas tight.
- No maintenance of the cell is needed.
- Can be recharged by constant current chargers.
- Is not damaged by a limited number of polarity reversals.
- Can be stored in a discharged state and is not damaged by total discharge.
- Fully charged cells can withstand continuous overcharge current indefinitely but are temperature dependent.

Disadvantages

- Has a limited life of 4 to 7 years. These are normally changed in the luminaire after 4 years' use.
- Cannot be recharged rapidly.
- Expensive in relation to its capacity.
- Has temperature limitations and charging problems at low temperatures under 5°C.

These cells are readily purchased as replacement items usually in D sticks and packs. They are welded in series and shrink/sleeved or wrapped into stacks or side by side. They are specified typically as in Table 7.1.

The actual specifications will vary between the different suppliers but the packs generally have either solder tags or may come with a two-way plug on a flying lead. They can be constant current charged at a non-continuous charge rate (cyclic) of 400 mA or continuous charge rate (float) of 200 mA.

Table 7.1 *Emergency battery backup packs*

Type	*Sticks*					*Side by side*
	1 cell	*2 cell*	*3 cell*	*4 cell*	*5 cell*	*3 cell*
Voltage	1.2	2.4	3.6	4.8	6	3.6
Nominal capacity	4 Ah	4 Ah	4 Ah	4 Ah	4 Ah	4 Ah
Length (mm)	59.5	119	179	239	299	63
Diameter (mm)	32.5	32.5	32.5	32.5	32.5	–
Width (mm)	–	–	–	–	–	103
Depth (mm)	–	–	–	–	–	36

Sealed lead–acid cells

These are available in many ampere hour ratings but are more limited in their working temperature range than NiCads. They are compact, maintenance free, reliable in operation and inexpensive but are damaged by exposure to high temperatures or deep discharge. Systems employing these cells must be fitted with low voltage disconnection and a temperature compensated charger to minimize battery capacity deterioration.

The sealed lead–acid battery used in self-contained emergency lighting systems uses a gel electrolyte and positive and negative lead electrodes. They have a design life of 4 to 5 years but in that time need no maintenance. Certain advantages and disadvantages exist.

Advantages

- Available in many ampere hour capacities up to 65 Ah.
- No maintenance.
- Readily available.
- Inexpensive in relation to their capacity.

Disadvantages

- Are damaged if fully discharged. Electronic circuitry within the luminaire should be incorporated to protect against deep discharge.
- Loss of capacity in relation to age.
- Must not be left in a state of discharge.
- Limited shelf life.
- If in a confined space adequate ventilation must be provided.
- Must not be placed in an atmosphere of, or in contact with, organic solvents or adhesive materials.
- Not to be used outside of the limits –15°C to 50°C for float/standby applications. This is the method in which the battery and load are in parallel to a float charger or rectifier so that the constant voltage is applied to the battery to keep it fully charged and to supply power to the load without interruption or load variation.
- Must be fastened securely if vibration is expected.

Sealed lead–acid batteries are typically as shown in Table 7.2 up to a capacity of 24 Ah and 12 V.

The use of sealed lead–acid batteries in self-contained lighting is limited to those that operate at a moderate temperature such as twin spotlights. As these batteries are used to support high light outputs the typical weight of the unit has been included in the table because

Table 7.2 *Sealed lead–acid batteries data*

Weight	*Nominal capacity*		*Dimensions (mm)*		
(kg)	*(Ah)*		*L*	*W*	*H*
0.31	1.2	6 V	97	25	55
0.57	2.8	6 V	134	34	64
0.87	4	6 V	70	478	105
1.32	7	6 V	151	34	97
1.98	10	6 V	151	50	97
0.58	1.2	12 V	97	48	54
0.7	2	12 V	150	20	89
0.83	2.1	12 V	178	34	64
0.95	2.3	12 V	178	34	64
1.12	2.8	12 V	134	67	64
1.2	3.2	12 V	134	67	64
1.75	4	12 V	90	70	106
2.65	7	12 V	151	65	97
4.1	12	12 V	151	98	97
6.1	17	12 V	181	76	167
9	24	12 V	175	166	125

this is a consideration in the mounting of the luminaire. If it is intended to use two or more battery groups in parallel to raise the system voltage they must be connected to the load through lengths of wire that have the same loop line resistance as each other. In addition it is not advisable to solder the terminals and it is a requirement to provide 5–10 mm free space between the batteries.

7.2 Central battery systems

There are a number of different battery types that can be found in these systems but the most popular are shown in Table 7.3. In general the nominal voltages are 24, 50, 110 and 240 V in the larger installations. Although the lower voltage levels are more economical the higher voltages are more appropriate in the larger site as they are more efficient in the containment of current levels and voltage drops.

It will be noted that the rechargeable batteries for central systems are essentially one of two types, namely lead–acid or nickel–cadmium. Both versions are available either sealed or vented. These batteries together with the cubicle are of necessity large and often need a dedicated room or area for housing which is in an area of low fire risk and away from electrical control gear switchgear and distribution boards.

Table 7.3 *Central battery system types*

Battery type	*Standard*	*Design life*	*Maintenance interval (months)*
Valve regulated gas recombination. Lead–acid	BS 6290 Pt 4	10 years	None
Valve regulated sealed Lead–acid	BS EN 61056	4–5 years	None
Vented lead–acid. Pasted plate	BS 6290 Pt 3	10 years	24–36 months
Planté. Lead–acid	BS 6290 Pt 2	25 years	6–9 months
Vented nickel–cadmium	BS EN 60623	25 years	12 months

All vented cells require a maintenance programme and may need to be topped up with distilled water periodically. As a cell discharges, the electrolyte weakens and the specific gravity (SG) falls until it eventually reaches an extent to which it can no longer provide energy. The state of the electrical charge of the battery at any time can be measured by the SG of the electrolyte and for this purpose a hydrometer is used. In practice the hydrometer draws an amount of electrolyte into a form of syringe and using a float technique the reading for the SG can be established in the syringe holder. The specific gravity is then related to a particular table to establish the percentage of charge of the cell. Although sealed batteries need no maintenance as such there is a need to ensure that their general condition is satisfactory and that the connections have not deteriorated or this will cause high contact resistance.

Sealed lead–acid batteries are regarded as being maintenance free and it is not possible to check the internal battery condition in a visual way. If any exact performance test needs to be carried out the only procedure capable of verifying the battery capacity is to perform a rated discharge test.

Those systems using valve regulated lead–acid types or gas recombination batteries need to have protection systems to prevent deep discharge or battery disconnection in the event of a long-term mains failure.

In the case of vented cells a boost or float charger is used as this provides a higher initial rate of charge but then gives a trickle charge for standby use. Valve regulated sealed lead–acid cells are charged by a temperature stabilized float charger which is a constant voltage type and regulates the battery charge current putting high currents into a discharged cell but low currents into a fully charged battery.

It is to be noted that if a high potential difference with cells is needed then the cells are to be connected to each other in series. If cells are connected to each other in parallel the potential difference will be the same as that for one cell but as a package they are able to provide greater currents. This is because their resultant internal resistance is less than that of an individual cell.

It is to be known that the vented batteries of lead–acid pasted plate, Planté and nickel–cadmium do emit potentially explosive gases when charging and consideration must be given to ventilation of the rooms in which they are sited. There are a number of factors that can be recorded for the essential battery types.

Valve regulated gas recombination. Lead acid

These need minimal maintenance and are of low cost with a life expectancy of up to ten years and capacities up to 400 Ah. They lose capacity with age and after 5 years have lost 20% of their efficiency. They are not to be left in a discharged state and have a limited shelf life off-charge that can be as little as 6 months. As with all lead–acid batteries they are not to be used in high ambient temperatures. They require little ventilation under normal operating conditions. The battery has a non-return control valve that allows the escape of gas if the internal pressure exceeds a given threshold. This valve does not allow air into the cell and no access to the electrolyte exists.

Valve regulated sealed. Lead–acid

These have been considered as an element in self-contained luminaires. They perform similar to the gas recombination type but have a lower life span and capacities up to 65 Ah. The maintenance cost of this battery is lower than that of other conventional lead–acid batteries up to its design life of 4–5 years. These also need little ventilation under normal operating conditions.

Vented lead–acid. Pasted plate

These are compact and inexpensive in relation to the Planté but have a shorter life span. Their plates are manufactured in the form of a grid and only a short initial charge is needed for the battery to be ready for use; however, their capacity reduces in relation to age and they are damaged if left in an uncharged state. This battery type makes good use of limited space and as they hold a large electrolyte capacity, maintenance is kept to a reasonable level, and they satisfy a realistic working service life span.

Planté. Lead–acid

These are high performance and have a long life expectancy and a capacity that is not significantly affected by age. Their cell condition can be checked by inspection. They are relatively expensive and are large in relation to their capacity. The Planté has a large electrolyte reservoir and therefore needs only modest maintenance. They are used extensively for standby installations where a long life is needed with little maintenance planning.

Vented nickel–cadmium

Like the Planté these have a long life expectancy with a good performance and reliability that is not easily affected by the working ambient temperature. They are not damaged if left discharged and in practice are found to have an indefinite shelf life in either a charged or discharged state. These batteries only need simple routine maintenance. They are relatively expensive and are not as quickly charged as many other battery types but are extremely robust and can accept high charge currents although they need a boost charge to achieve full recharge.

The manufacturers of the batteries provide data to specify the battery nominal output volts, output amps and the output watts for 1, 2 and 3 hour durations. An example appears in Table 10.1.

7.3 Maintenance – battery rooms and ventilation

Maintenance

Although the manufacturers' data sheets will always contain advice on care and maintenance of their batteries there are a number of time honoured general procedures that apply:

- Only carry out maintenance in a well-ventilated area.
- Never allow the level of the electrolyte to fall below the tops of the plates. In the event that a loss of electrolyte has occurred because of evaporation this can be replenished by adding distilled water.
- Do not allow the use of naked flames in proximity to any battery cells.
- If removing a cell from use for any period of time it is first to be fully charged and the electrolyte is not to be removed. A periodic charge can be made to the battery until it is required for further use. As a rule it should not be stored for longer than 8 weeks before undergoing a recharge although the actual shelf life between the battery types varies.

- Do not store cells uncharged as sulphation of the plates will occur and the capacity of the battery will be reduced because of the internal cell capacity increasing.
- Coat the terminals of batteries with petroleum jelly to stop corrosion occurring.
- Batteries have the potential to deliver high dc currents and because their very construction inhibits them from being conventionally isolated care must be exercised in handling and storing. They are always to be considered as being electrically live.
- At the end of their life batteries are to be disposed of according to the manufacturers' instructions.

Battery rooms and ventilation

The batteries are themselves housed in heavy gauge steel cabinets with the vented cells tiered on acid resistant steel trays for easy maintenance. The central battery cabinets, depending on the size of the system and the physical dimensions of the total battery network and control gear, will vary. The cabinet may be wall mounted or floor standing on a plinth to prevent corrosion or moisture damage. These will all have easy access to the cells via hinged lockable doors. In every application these are to be in a secure and safe environment and a unique room may be needed to satisfy this end.

Vented batteries when charging do emit potentially explosive gases and although valve regulated batteries need little ventilation under normal circumstances there is still always provision to make for risks as a result of fault conditions. In order that gases can be purged from the room an adequate level of ventilation is required. This is generally satisfied by the use of a low level intake supplemented by high level vents. When specifying the area it is advisable to budget for boost charging conditions.

The formula used to determine the number of air changes per hour (A) is

$$A = \frac{0.045 \times \text{number of cells in the battery} \times \text{charge rate in amperes}}{\text{Room volume (cubic metres)}}$$

This determines the number of air changes per hour during boosting. In practice the system batteries are on float charge for most of their working life and rarely on boost. Nevertheless it is normal to account for the boost conditions when establishing the ventilation needs although the amount of gas actually liberated on float charge is only some 1.5% of that generated on boost.

There are certain other considerations of every application:

- Floors are to be flat, vibration free and robust.
- Access is to be granted to authorized persons only so the room and cubicle should be locked.
- A free area should exist around the cubicle to be used as a working area.
- Notices are to be displayed on the entry door – 'Battery room', 'No Smoking', 'Extinguish all naked lights before entering'.
- The battery room is to be well illuminated and dry and be of an ambient temperature in the order of 10°C–27°C.
- The area should have safety equipment with first aid facilities because batteries can be of an explosive source and cause chemical burns. An eyewash station should be provided.

8 Low mounted way guidance systems and ancillary applications

As with any electrical network there are always components that can be added to the system to extend its role or to customize the way in which it is applied. This is necessary in order to satisfy more diverse needs of the system for specific duties and options. In the second part of this chapter we look at a number of components that can be used to this end. It may also be noted that to this stage we have investigated the conventional form of emergency lighting. However, in those emergency situations where smoke can be a distinct hazard there is a need to consider optional low mounted way guidance systems that are installed at or close to the floor level. This is because under fire conditions hot smoke rises rapidly to form a layer from the ceiling downwards leading to a reduction of visibility at overhead positions.

8.1 Low mounted way guidance systems

In fire situations our reactions are to stay close to the floor when attempting to escape because at that point the air is cooler and the smoke is less dense and toxic. Therefore we have an argument for the installation of low mounted emergency lighting systems to provide a highly visible line of sight at or close to the floor level. This is because overhead luminaires would be further obscured by the smoke rising to the ceiling level and then drifting downwards. Research suggests that low mounted way guidance systems have particular benefits for those applications of particular fire risk and for those buildings catering for the elderly and visually impaired people. As an overall concept it has been found that as an ideal a combination of both conventional networks and low mounted way guidance systems was favoured. If serious fire risks exist and these ask for the complete shutdown of all electrical systems so that the emergency lighting can provide the only illumination technique, these way guidance systems are of particular value for the emergency services. They also have purpose designed signs and indicators to further improve the information passed to evacuees in the event of an emergency. This is to provide directional information, clear identification of escape doors and the location of fire fighting and safety equipment.

The exact standards in which we are interested with respect to these low mounted way guidance systems are BS 5266 Part 2 and EN 60598-2.

In practice lengths of surface or flush mounted modular construction profiles are installed at floor level or along walls to follow contours of the building materials. They are effectively long strips that are used to house incandescent lamps or high intensity LEDs so that an optimum number of lamps is held in any fixed span. These will give high intensity leading to good levels of visibility even in those environments subject to smoke although they can also be advocated for use in clear conditions. In the installation itself the modules are bridged to each other by connectors with special socket connections in a parallel ring circuit which is so arranged that a break at any point does not render the system inoperative.

Figure 8.1 illustrates the final installation technique.

The advantages and safety benefits can be endorsed as:

- Highly visible even in areas subject to dense smoke.
- Provides the emergency services with speed of access.
- Improves evacuation times.
- Aids safety especially for the partially sighted.
- Easily installed.
- Low power consumption.

Figure 8.1 *Low mounted way guidance systems*

The exit or escape route directional indicators which are needed as part of the system either plug into the guidance way circuit or clip into the modular construction profiles in the same way as the lamps to form a neat and simple technique of showing emergency routes to be followed. There is a wide available range of high luminous output legends and exit indicators so that they are still visible in smoke conditions.

It is important to note that the indicators used with low mounted way guidance systems should be used in conjunction with conventional high level signage to achieve full compliance with appropriate regulations.

The main power supply unit provides the mains supply to the system. It incorporates the necessary circuitry, charger and the maintenance-free sealed lead–acid batteries to provide the standby power source in the event that the mains supply fails. It can be additionally energized in response to a fire alarm generating a signal. These power supplies may be mounted at strategic locations in the protected premises to ensure that local subcircuit monitoring is achieved. Conventionally they may be non-maintained or maintained with a standby of up to 3 hours' duration. Maintained versions may also be used to provide energy efficient night lighting and security lighting for applications in premises such as nursing homes and hospitals.

The prominent and established low mounted way guidance systems presently in use offer incandescent lamp or LED modules. However, with progression in fibre optic techniques in mainstream lighting networks using an ability to transmit light along fibres by internal reflection to a wide variety of precise points, we can expect developments of these in emergency lighting concepts.

Fibre optics are ideal for general low level guidance illumination and are now used for innovative projects for highlighting architectural features at night and for those areas where a high level of vandal resistance is required. No heat or UV is present in the light beams so lighting is safe and precise.

Low mounted way guidance systems can be used in conjunction with conventional emergency lighting luminaires by linking the disciplines. Indeed as with any electrical and building management systems there are techniques available to enable one system to complement another by using integrated technologies. For all emergency lighting systems there will be ancillary equipment that can be sourced to further improve the capacity of the installation. In the case of specialist equipment the manufacturers will provide optional components to expand the system or customize it. For conventional emergency lighting applications many items of ancillary equipment are already subject to widespread use.

8.2 Ancillary equipment and duties

Emergency lighting can be extended in its duty or be customized by the addition of optional extras and ancillary equipment. These systems can also be given a measure of integration with other electrical services by the connection of additional devices. It is often the case that even more sophisticated monitoring must be carried out or alarm units need to be incorporated so that warning of a fault or failure of a function has occurred. There are a number of devices available that can be considered to fulfil ancillary requirements. The most widely accepted applications follow:

- *Auxiliary alarm relay.* This is typically a relay with changeover switching contacts that can provide a remote alarm indication for a specified event. A signal from this event is used to trigger the relay coil and the main contacts can then be used to switch on either an audible remote device or a visual indicator. Can function as a remote alarm unit with a sounder mute facility to operate as a simple fault report giving a visible and audible warning in a manned operations room.
- *Earth fault detection monitor.* Detects leakage currents in excess of 10 mA in either the positive or negative poles to trigger a fault.
- *Fire alarm relay.* This is a specific relay that is energized by 24 Vdc from the fire detection system in the same building. In the event that the fire alarm has been activated the hold-off supply from the continuously rated relay is removed and the emergency lighting is brought into effect. This is often used in conjunction with low mounted way guidance systems but can be used with conventional networks.
- *Hold-off relay monitor.* Used in load switching. It holds off the central battery unit maintained output and allows for non-maintained lighting operation with the normal final lighting circuit being monitored. Figure 8.2 shows a typical arrangement.
- *Night switch.* A switch device usually in the form of a key to prevent any battery discharge for premises unoccupied at night.
- *Solar clock.* Monitors the light levels at sunrise and sunset and automatically controls the maintained output.
- *Subcircuit monitor.* Used in non-load switching to interrogate the normal mains lighting. It makes a signal available to the central battery unit if the final lighting circuit fails. Figure 8.3 shows a typical arrangement.
- *Three phase relay.* This is used to monitor all of the phases and will activate the emergency lighting if any of the phases is disconnected at the cubicle. By monitoring all of the phases the emergency lighting can be brought into effect if any of the phases fails.

- *Timer (24 hour/7 day)*. Used for maintained systems the pre-programmed on/off sequences can control the maintained output.
- *Timer boost*. Automatically provides a boost charge via a timer for a specific period during the battery recharge cycle. It comes complete with a manual override.
- *Timer delay test*. Used to simulate a mains failure. The timer is automatically reset on completion of the test and the batteries are then recharged.
- *Timer run on*. This maintains the emergency lighting in an energized state for a given period in the order of 10 minutes when the mains has been restored.
- *Tripping units*. Switch tripping units are used to provide power as needed for trippng switchgear contactors and operating small motor mechanisms plus indicators.
- *Vented cells – low electrolyte*. Indicate a low level by means of a probe inserted into a sample cell.

All of the aforementioned devices are basic components that may be used with virtually any system.

It will be found that some are specified as standard equipment within the instrumentation and control gear whilst others may be provided as optional extras to complement a system.

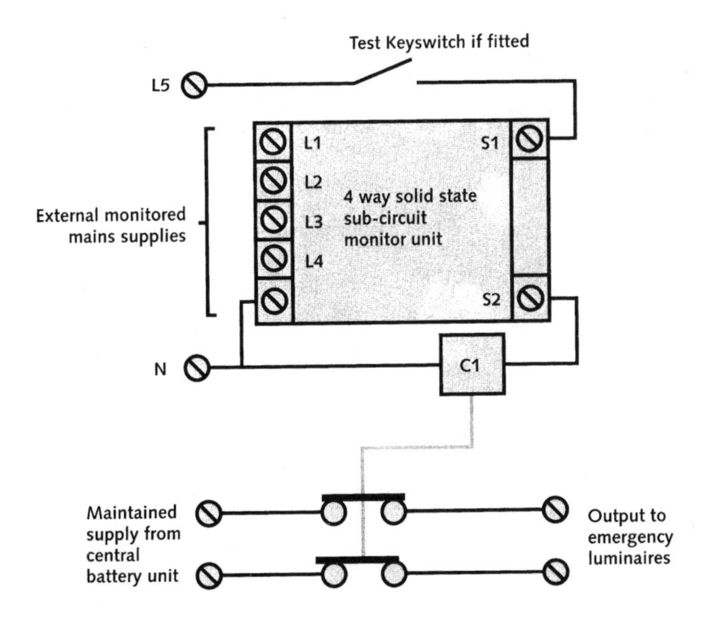

Figure 8.2 *Hold-off relay arrangement*

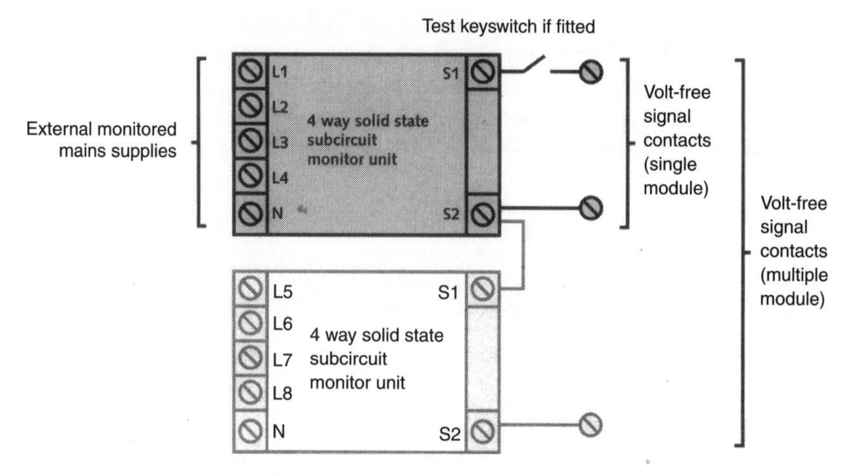

Figure 8.3 *Subcircuit monitor*

The components that are used specifically in emergency lighting for automatic test purposes and intelligent testing practices are given special identities as shown in the next chapter.

9 Automatic intelligent testing

The testing and maintenance of emergency luminaires is vital to ensure ongoing compliance with current legislation. For those buildings with fire certificates evidence is actually needed to verify the performance of the emergency lighting. Indeed further emphasis is being put onto the testing of the systems and evidence of appropriate testing is an ongoing requirement.

Those areas that do not specifically need fire certificates are still governed by the Fire Precautions (Workplace) Regulations which demands that they are maintained and kept in good repair. In order that the testing and maintenance is correctly performed it is normal to elect a responsible person to keep records of the system installation details plus the test reports and the maintenance and action taken as a result of these activities. In general, the routine tests are performed by the occupier of the premises although it is at times more appropriate to have these practices carried out by a third party. The results of all the testing and corrective action must always be kept in a log and such records are to be available for inspection by an authorized person. Test record forms are supplied with most luminaires to aid this process.

The intention of implementing intelligent testing is to ensure that the testing cannot be overlooked as it is done to an automatic routine perhaps also with remote monitoring and support diagnostics by a PC. The correct testing in the traditional way has always been a particular concern in those buildings and in business premises that have limited qualified personnel. Indeed traditional manual-based testing can be expensive in terms of man-hours. There are also those premises in which it is impractical to perform entire system tests at the same point in time. As an example hospitals need individual group testing for safety but this leads to a more complex and time-consuming exercise. This results in a greater maintenance cost and more inconvenience. We have previously considered the testing and maintenance of self-contained systems in Chapter 4, section 4.4, and it will be noted that this is demanding as it extends to daily, monthly, 6 monthly and 3 yearly programmes.

For central battery systems the tests in which we have an interest are as follows:

- *Daily*. Visually check all luminaires and confirm that all maintained lamps are operating correctly. Report defects for attention. For central power supplies check that indicators are healthy.
- *Monthly*. Check that all luminaires are clean and in good condition. Ensure that fluorescent lamps are bright and do not suffer from blackening or flickering. Simulate a mains failure for a time sufficient only to verify the emergency operation. For central power supplies this is to be carried out by means of the control switch at a time it is safe and convenient to do so.
- *Six monthly*. The tests are to be as those for monthly but for one-third of the rated duration at a time that is safe and convenient to do so.
- *Annually*. The tests are to be the same as monthly but for the full duration at a time that is safe and convenient to do so. The test is not to be done when the building is occupied because the batteries will take up to some 24 hours to recharge.

For all of the tests a record must be taken of the date and results and these are to be held in an emergency lighting log book. Faults are to be corrected and the date of their completion recorded.

In order to ensure that systems of both types remain in full working order manufacturers promote service and maintenance contracts to the users of their equipment. Under such schemes visits will be made to the particular premises at an agreed given frequency normally twice a year for self-contained systems and quarterly for central battery systems. This may well contain a contract covering the full cost of parts and labour together with a 24 hour emergency call-out service. An option to this is to install automatic intelligent testing. The methods vary between the different goods and from the various manufacturers but are always either self-testing (self-check) or centrally addressable. It is to be noted that some intelligent systems are for use with self-contained luminaires and others are intended for central battery lighting networks. There is an overlap of terms but self-testing infers a system using self-contained luminaires. Centrally addressable is a term often used with both self-contained and central battery systems.

Self-testing systems are self-checking and intended for self-contained luminaires that are installed in small to medium size premises. They need no additional wiring run to the emergency luminaires so are easily added to an existing installation and therefore cause no disruption to the premises or décor. They employ an internal microprocessor/module that is added to the luminaire and has a pre-programmed test sequence. The status of the module can be controlled by an infrared link to a hand-held controller with an event display for viewing or perhaps downloading to a PC. The selection of the exact test with regard to duration or self-testing can be carried out from such a controller. The module within the luminaire contains LEDs to show the light fittings' full status. It is interrogated by aiming the IR hand-held controller at the module from a distance of up to some 5 metres and pressing the appropriate buttons on the handset to initiate a test sequence. These devices are simple to use and can be operated by most members of staff. They are actively promoted for those premises that require a fire certificate and where the test parameters may vary. Infrared programming enhances the commissioning of larger systems as set up is achieved without having to gain access within the luminaire for programme amendments so a saving in man-hours is achieved. Systems tested by this nature are often also called semi-automatic testing systems.

An alternative form of module is offered by certain manufacturers for fitting within the luminaire housing or in the remote box of an emergency conversion kit. It is low cost and simple to set up and promoted for small systems. It also carries out tests automatically with any faults detected being indicated by a sequence of flashes on the modules' integral LED or by an audible signal. The tests are performed to a cycle so that they are clearly defined in terms of times and that is initially made selectable by means of a series of DIP switches built into the module. These determine an offset of time and day from power-up/commissioning to test. It therefore has a fixed time regime with manual set-up. In addition the diagnostics continuously monitor the battery voltage and charge current. The diagnostic unit can be ordered as 'in-built' with the luminaire at the purchasing stage. These systems ensure that automatic tests are performed at a convenient and defined time aiding maintenance programmes in the event that remedial action may need to take place.

For those self-contained systems that are greater in size, for example airports and hospitals and where there is a need for central control, monitoring and remote testing plus diagnostics with hard copies of test results, it is necessary to install addressable systems. These networks are referred to as fully automatic central testing systems as they report all faults back to a central control point that can be sited in a maintenance area used for other safety and security monitoring functions. The system is commissioned by fitting every luminaire with an interface module and an address number by which it is then identified. The actual wiring techniques differ but they tend to have a control network twin screened cable in addition to the mains cable that is connected in a loop and back to the control panel – see Figure 9.1.

The control cable specification type and gauge size is governed by the number of light fittings in the network and the total cable length.

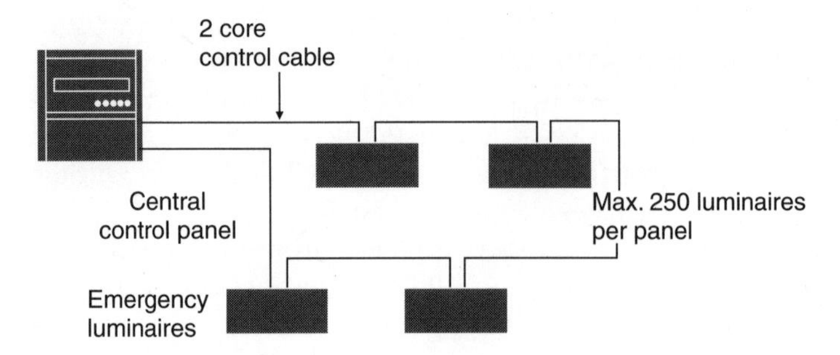

Figure 9.1 *Addressable network*

In some systems an end of line resistor can be found at the luminaire position furthest from the central control unit. In practice each luminaire will have a mains supply run to it together with the control network loop. On commissioning the control panel analyses the luminaires in terms of location and type and the information is formatted into PC software. A test regime may then be compiled and carried out automatically with appropriate test periods. The central control panel will display all of the system information that can then be downloaded to a printer for hard copies. In effect the system can be configured from a control panel with an appropriate display for events or via a PC with a touch screen perhaps also with modem connection to allow remote testing and report retrieval from a central office.

With systems of this nature the luminaires can be allocated into groups pertaining to an area for test purposes or they may be tested individually. In addition the full duration test is in a position to be automatically rescheduled if a luminaire has been in emergency operation recently and this will allow a full recharge first.

It is necessary to programme the system so that sector testing is employed to ensure that the whole network is not tested simultaneously. An option can be to switch maintained luminaires on and off from the control unit to include exit signs. This is useful in public buildings with multiple maintained lights that are vacated in the evening as they can then be switched off from the panel. User and service security codes can form a part of the system so that before any manual test or programming function can be undertaken an authorized PIN must be entered. Features such as voltage-free auxiliary contacts in the panel allow remote indicators to be easily added and it is often the case that slave panels or displays can be installed at additional locations. For verification purposes printers can produce evidence that the full emergency lighting has been tested and whether this has been done automatically or manually together with the time and dates of the test results. It is important to recall that the functions that need to be checked are:

- mains present and healthy;
- battery present;
- battery in circuit and charging;
- inverter circuit in emergency operation;
- lamp functions and in circuit – duration as necessary.

It will be found that certain components used in addressable emergency lighting systems and applied to self-contained luminaires are also used to some extent in central battery systems although the type of hard wiring used and the techniques by which they are applied differ.

Particular automatic testing systems are available with any central battery system and include central inverter types. They are promoted where regular, routine, manual testing would be time consuming, disruptive and inconvenient. They may be used to implement weekly automatic short duration testing of the fittings and the emergency power source. In addition they may have a facility to periodically instigate full load testing of the entire network. The full load duration testing may be called for by a manual means or pre-set for automatic selection of 1, 2 or 3 hours' duration. In the event that the systems low voltage circuit operates before the duration test is concluded the length of discharge achieved is recorded. Slave luminaires may be fitted with a photocell to measure the lamp output during the test and this reading is compared to a pre-set factor. If the reading is less than some 80% of the desired value a fault is reported at the control panel. The exact geographical location of the fault can be recorded and displayed on the screen or on an optional printout. In those cases where it is not possible to install a photocell a remote current measuring shunt can be used.

The slave luminaires are assigned to a number of available time clocks that determine the time and day of their actual test regime. Any luminaire can be assigned to any individual time clock or a combination of time clocks as this enables a wide range of automatic rest schedules to be established including those specified in British and European standards. The system therefore comprises a central control panel, optional printer, communication cable, luminaire interfaces, zone switching interfaces, photocells and the central battery unit interfaces. Network monitoring panels can also be added if there is a need to have multiple panels in use on one site. When used on maintained systems any interruption of the 24 Vdc test supply can be adopted to bring on all of the emergency lamps. This can be used to great effect to switch on all of the emergency lighting in the event of a fire alarm signal or security alert. Fail safe communications via a dedicated data line allows the system to be used integral with fire alarms and other security systems. In practice even when used with extremely complex emergency lighting networks a single relay interface can be used to ensure that all emergency fittings connected to a given panel are brought into effect if a fire alarm activation is recorded. The fully addressable interfaces also enable specific luminaire location and replacement lamp information to be displayed so that every luminaire is separately accountable.

Automatic testing systems are designed to function with almost any type of luminaire by measuring its light output and to ensure user friendliness indicators are provided to guide system users through the various menus and programming options. The panel itself requires little other than a data line connection and mains supply. The data cable that is to be used with the different systems is as specified by the particular manufacturer. This cable will be governed by the wiring runs and number of luminaires per circuit. Account must also be made of the power consumption of each address interface. An indication of a typical system is shown in Figure 9.2. This illustrates dedicated interfaces within each luminaire plus additional interfaces to control each external hold-off or changeover device.

Automatic testing is always best installed from the outset as testing for the future is readily controlled and a history of the system can be compiled for effective maintenance planning. In addition the person carrying out the testing needs only basic training and even with the

Dedicated interfaces within each luminaire plus additional interfaces to control each external hold off or changeover device.

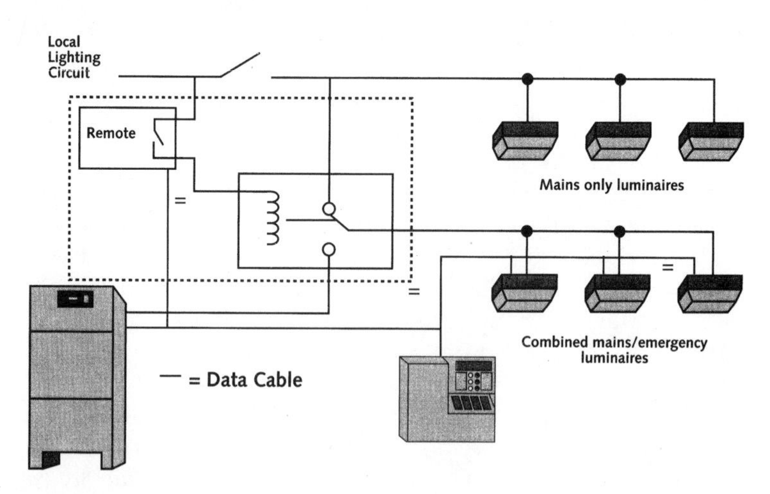

Figure 9.2 *Wiring arrangement – automatic central battery testing system*

more complex central battery system the operator need only be familiar with general PC-based techniques. The testing therefore becomes flexible, low cost and full fault diagnosis and reporting are made available.

In the event that a system is already in operation and at a later stage automatic testing is wanted it may be achieved by the addition of a data cable and modification to the luminaires. If a small self-contained system is in existence and if the fabric of the building is to be undisturbed it may be reasonable to use a semi-automatic system to offer a self-test option so that all the lamps are tested on an individual basis. This is a rapid and simple operation.

There are a great many options on offer for testing enabling a user to run an effective and efficient test regime that conforms to the legal requirements although this will always be governed by the system to be employed and the level of investment that can be made.

10 Reference information

10.1 Definitions and terminology

Anti-panic (open) area lighting. A part of the emergency escape lighting which is installed to reduce panic and provide illumination and direction finding to enable people to reach escape routes safely.
Automatic testing. A programmable central test regime.
Axis. The distribution of light from a plane projecting from the longer side of the luminaire (transverse axis) and the shorter side (axial axis).
Ballast. A unit with a connection to the supply and one or more fluorescent lamps and which by means of inductance, capacitance or resistance serves to limit the current of the lamp or lamps to a required value. This may also initiate transforming from the supply voltage and preheating current, inhibit cold starting, correct the power factor and suppress radio interference. A ballast controls the operation of a fluorescent lamp from a specified low or high voltage ac or dc source typically between 12 and 240 volts. An emergency ballast is the module that operates the lamp in emergency mode.
Ballast factor lumen. The ratio of the light output of a lamp working from a ballast to the light output of the same lamp powered from a suitable reference ballast at its rated voltage and frequency.
Battery. Secondary cells that are intended to provide a specified level of power when the normal supply fails. These will be interconnected to form the standby power supply for an emergency lighting system.
Battery capacity. The discharge capability. This is a product of average current and time expressed as ampere hours (Ah) over a given stated period of time. A shorter total discharge period gives rise to a smaller available capacity. A derating factor may be applied if the duration is less than the rated time.
Central battery system. A system employing a battery and related control circuits to enable emergency power to be generated to slave lights. The batteries for a number of luminaires are housed in one location, sometimes for all of the luminaires in a complete building but generally for the lighting units on one subcircuit.

Central inverter. An inverter for more than two lamps operating luminaires sited at a remote point. The output voltage and frequency waveform may differ from that of the normal mains supply.
Certification. The industry standard for the manufacture of emergency lighting equipment and monitored by the appropriate approval authority.
Changeover device. A device intended to connect the battery by an automatic means to the emergency lighting circuit in the event of the mains supply failing.
Charger (battery). The component of the control unit that provides the charging to the battery cells from the normal supply.
Combined (sustained) emergency lighting. A luminaire with two or more lamps. One of these is energized from the battery with the remainder energized from the normal supply. In practice this type of luminaire can operate at all material times.
Continuous operation system. A circuit that operates for an indefinite period of time such as those used for maintained lighting.
Conversion module. A mains luminaire converted to operate in emergency mode.
Correction factor. A factor introduced to allow for light reduction due to the effects of age, dirt etc. See also **Service factor**.
Design voltage. The specified operating voltage. This is declared by the manufacturer to which all of the ballast characteristics are related. This is 80% to 90% of the maximum value of the rated voltage range.
Designation. BS EN 60598-2-22 Annex B code for type, category, mode and duration.
Disability glare. Effects of light preventing the route being seen clearly.
Diversity ratio. A ratio of minimum to maximum illuminance at the working place.
Duration. Governing the mains failure period.
Emergency escape lighting. The part of an emergency lighting network which is included to ensure that the means of escape can be safely used at all material times.
Emergency exit. An exit primarily for use during an emergency only.
Emergency lighting. That lighting intended for use in the event that the normal building supply lighting fails.
Escape route. A route that forms part of the means of escape from a point in a building to a final exit.
Exit. A way out of the premises that is intended for normal use when the premises are in general occupation.
Final exit. The last opening to the outside of an escape route. This exit is to be one in which there is no risk to life directly from a fire or hazard that is present within the building.

Fire certificate. A document issued by the fire authorities in compliance with the Fire Precautions Act as applicable.
Fire exit. An exit intended for use only in the event of a fire.
'F' mark. An indication that a luminaire may be installed on a flammable surface as it is unable to ignite that surface.
High risk task area lighting. That illumination intended for potentially dangerous work areas. This is intended to safeguard people involved in dangerous processes or situations and to allow clearly defined shutdown procedures to be carried out. It is to ensure the safety of operators and other building occupants.
ICEL. Industry Committee for Emergency Lighting.
ICEL 1001. The registration of the industry standard for the approval of photometric performance and claimed data of emergency lighting equipment as tested by the British Standards Institution (BSI) in the UK.
Illuminance. The measure of luminous flux density or the luminous flux incident per unit area at a surface. This is measured in lux (lumens per metre squared). It is accepted as the general term for light.
Inhibit mode. A control mode used to inhibit the discharge of the emergency lighting batteries at times when the building is unoccupied. If a mains failure occurs at this time the batteries are fully charged so the building may be occupied when required. It is performed by an inhibit switch that must be interlocked with the essential building services so that the premises cannot be inadvertently occupied without the emergency lighting being recommissioned.
Internally illuminated sign. A sign complete with an integral lamp mounted internally.
Inverter. A device intended for the conversion of dc to ac power.
K factor. The ratio between the light output from a lamp in its worst condition to the light output at nominal voltage. This worst condition is typically at the end of discharge and with any particular voltage drop.
Life (maintained) factor. A correction factor for maintained lamps to allow for reduced light output through the lamp life.
Light output ratio (LOR). The ratio of the total luminuous flux from a luminaire to that of the sum of the individual light outputs of the lamps operating outside of the luminaire under reference conditions.
Lighting design lumens. The quoted nominal value and use to represent the average output of the working life of a lamp.
Lumen. SI unit of luminous flux density.
Luminaire. Apparatus which distributes, filters and transforms the lighting given by a lamp or lamps and which includes all of the items necessary for fixing and protecting these lamps and for connecting them to the supply circuit. Internally illuminated signs are a special form of luminaire.
Luminance. The measure of light reflected from a surface.

Luminosity. The apparent brightness of light from a source.
Lux. The level of illumination produced by 1 lumen over 1 square metre.
Maintained emergency lighting. A system in which the emergency lamp is in operation at all material times being powered from the ac supply under normal circumstances and from a battery in the event of mains failure.
Mode. The luminaire operating status.
Mounting height. The vertical distance between the luminaire and the floor.
Non-maintained emergency lighting. A system in which the emergency lamp operates only in the event of mains failure.
Normal lighting. The permanent artificial lighting to supplement natural light during daytime or for illumination purposes during the hours of darkness.
Open area. A route or area that leads to a designated escape route.
Phase failure detection. A device intended to monitor a supply phase that provides normal lighting and is intended to bring into effect appropriate emergency lighting if that phase fails.
Pictogram. The legend in a graphic form.
Rated duration (battery). The declared period of time a battery will continue to support the load. This is generally 1 or 3 hours of at least 85% of its nominal output voltage.
Rated load. Maximum load that can be connected to a battery for the rated duration.
Recharge period. The time necessary for the batteries to regain adequate capacity of charge to enable the lamp to perform its rated duration.
Rest mode. A control mode enabling the emergency lighting to be switched off during a mains failure and at a time when the premises have been fully vacated. This is an aid to reoccupation of the building because it can stop full discharge of the batteries.
Risk assessment. The technique applied to assessing hazards and risks based on chances of harm arising from a hazard.
Room index. The relationship between the dimensions of a room in terms of height, width and depth for use in those calculations used in illuminance calculations. This is required in order to determine the luminous flux incident.
Sealed battery. A cell that is so sealed in its construction that no provision is made for the replenishment of electrolyte. These may include a safety venting system. They may also be referred to as recombination.
Self-contained emergency luminaire. A luminaire containing a battery, lamp and all of the control circuitry within its enclosure or within a contained enclosure adjacent to and at a maximum distance

of 1 m from the main housing. The luminaire may provide maintained or non-maintained emergency lighting. These are also referred to as single point.

Self-contained module. An apparatus containing all of the components needed to operate an emergency lighting luminaire. This tends to comprise the battery charging circuit, the dc to ac inverter and the changeover device.

Semi-automatic testing. A test for emergency luminaires conducted on a local basis.

Service (maintenance) factor. A calculation to allow for the effects of ageing and accumulation of dirt. It is the ratio of the illuminance provided by an aged system with dirty luminaires to the illuminance of the same system when new.

Single point luminaire. See **Self-contained emergency luminaire**.

Slave luminaire. A luminaire that is intended to be used with a central battery system so that it does not have its standby battery source integral with the enclosure. It may also be called a centrally supplied luminaire.

Spacing/height ratio. The ratio of spacing between the geometric centres of adjacent luminaires to their height above the floor.

Stand-by lighting. A part of emergency lighting which may be provided to enable normal activities to continue in the event of a mains failure.

Storey exit. Access to a protected stairway.

Sustained emergency lighting. See **Combined emergency lighting**.

Transistorized ballast. An inverter/slave ballast. A dc to ac inverter using transistors for supplying power to one or more fluorescent lamps. They may be dc only or ac/dc working.

Transitional emergency lighting system. Emergency lighting intended to operate for an interim period.

Uniformity ratio. The relationship of average illuminance to minimum illuminance measured at the working surface.

Unsealed battery (vented). A battery that requires electrolyte replacement that can be made to a maintenance programme. Vents within the enclosure allow the vapour products of electrolysis to escape.

Utilization factor. At zero reflectance it determines the proportion of light output from a lamp incident direct to the floor for differing room indexes and diffusers.

10.2 Certification and standards

Throughout the developed world different countries have their own legislation which must be adhered to in order to produce a safe environment. It will be found that certain countries produce national legislation whereas in others the laws relating to health and safety may differ. In the UK it is the Health and Safety at Work Act that contains the duties for employers together with the general duties that employees have to themselves and to each other. This also sets out the responsibilities towards the general public and becomes a foundation for the principles of installing emergency lighting.

The registration and certification scheme for emergency lighting is administered by the Industry Committee for Emergency Lighting (ICEL). The ICEL schemes extend beyond those of the BSEN product standards as they include additional requirements for product performance including factors such as battery life, fire retardancy and component life expectancy.

For dedicated self-contained emergency lighting products it is normal to specify that they are Kitemarked and registered to ICEL 1001. This registration requires products to be Kitemarked and also requires independently verified photometric information.

For converted mains luminaires it is normal to specify that they are produced by an ICEL 1004 manufacturer and that the components used are Kitemarked. This ICEL 1004 registration ensures that the conversion process uses Kitemarked components, involves full electrical, thermal and EMC tests and that the work is performed within a BS EN ISO 9002 manufacturing environment. The format of the legends of internally illuminated exit signs are to comply with the Health and Safety (Safety Signs and Signals) Regulations 1996 or the European Signs Directive and the units themselves are to comply with the appropriate product standard.

Kitemarking demonstrates that independent tests have proved compliance with BS 4533 102.2.22 and that the manufacturing unit is a BS EN ISO 9002 registered site.

It is practice to use products that have been certified to a product standard since these are then third party verified as safe in use when installed as intended and they will fulfil their desired operational function. Safe in use implies that the product is not a shock or fire hazard and that non-operation is not a safety hazard in an emergency situation.

There are many laws and codes of practice that interpret and implement legislation requiring emergency lighting. It is normal to classify the areas for emergency lighting as:

(1) Areas which are generally exempt from legislation or the need for a fire certificate. These are typically private dwellings and workplaces used by the self-employed.
(2) Areas where legislation applies but fire certificates are not required. This covers many areas following a deregulation of premises

needing a fire certificate. The regulations in this case state 'Emergency routes and exits must be indicated by signs and emergency routes and exits requiring illumination shall be provided with emergency lighting of adequate intensity in case of failure of the normal lighting.' This is carried out by means of the owners or occupiers performing a fire precautions risk assessment. If five or more people are employed there is a legal requirement to record significant findings of this assessment and to document the measures taken to deal with those risks. If fewer than five people are employed the risk assessment and remedial action need still be carried out but not documented. The staff/representatives must be informed of the risk assessment findings and be given access to all documented information.

(3) Areas requiring fire certificates. The owner or the occupiers of these areas do not need to carry out a risk assessment since this is carried out by the inspecting body such as the local fire authority.

(4) Areas not requiring a fire certificate and not falling under the Fire Precautions (Workplace) Regulations 1997. These have legislation but specific to the requirements of those areas. These areas are mine shafts and mine galleries, and workplaces covered by a safety certificate issued by the Safety of Sports Grounds Act 1975 or the Fire Safety and Safety of Places of Sport Act 1987 when used for a purpose related to the certificate. Also in this category are subsurface railway stations, construction sites, ships within the Docks Regulations, certain means of transport, agricultural and forestry land away from the undertaking's main buildings and many offshore installations.

The risk assessment is a process to guide an assessor to identify the risks and the need for fire precautions in order to reduce the risks to an acceptable level. When reducing those risks the need for the emergency lighting can be determined. At the point of deciding to install an emergency lighting system the British Standards Codes of Practice for the emergency lighting of premises, BS 5266 Parts 1 and 7: 1999, are to be invoked and the emergency lighting luminaires should conform to the harmonized British and European product standard BS EN 60598-2-22. It was the Fire Precautions Act 1971 that began the need for employing emergency lighting and this was later amended by the Fire Precautions (Workplace) (Amendment) Regulations 1999 introducing the risk assessment method. It effectively places an emphasis of complying with the regulations on a responsible person such as the owner of the premises, an employer or an architect. It should be noted that even with a fire certificate a risk assessment is still a requirement, and that in any premises with 20 or more employees a fire certificate is needed to confirm that adequate fire precautions have been instigated.

There is a great deal of legislation and standards that govern the use of emergency lighting systems to include:

ICEL standards and guides

ICEL 1001. 1999: Scheme of product and authenticated photometric data registration for emergency luminaires and conversion modules.
ICEL 1003. Applications guide. Emergency lighting.
ICEL 1004. 1996: The use of emergency lighting modification units.
ICEL 1006. 1997: Emergency lighting guide.
ICEL 1008. 1998: Guide to risk assessment.
ICEL 1009. 2000: Emergency lighting central power supply system registration scheme.

Standards

BS 5255. Photometric data of luminaires.
BS 5266. Part 1: Code of practice for the emergency lighting of premises other than cinemas and certain other specified premises used for entertainment.
BS 5266. Part 2: Code of practice for electrical low mounted way guidance systems for emergency use.
BS 5266. Part 7 (BSEN 1838): Lighting applications – emergency lighting.
BS EN 60598-1: European standard for luminaires. Refer also to BS 4533.101.
BS EN 60598-2-22: Particular requirements – luminaires for emergency lighting. Refer also to BS 4533 102.2.22.
BS 5499: Specification for exit signs.
British Standard Code of Practice CP 1007: Maintained lighting for cinemas.
BS 764: Electromechanical relay contactors.
BS 5424. Part 1: Electromechanical relay contactors.
BS 6290. Parts 1–4: Lead acid stationary cells and batteries.

Regulations and bylaws

Fire Precautions Act 1971.
The Fire Precautions (Workplace) Regulations 1997, as amended by the Fire Precautions (Workplace) (Amendment) Regulations 1999.
Health and Safety (Safety Signs and Signals) Regulations.
Building Regulations, Approved Document B. 1991.

Health and Safety at Work Act 1974.
Cinematographic Act 1952.
Cinematographic (Safety) Regulations 1955 No. 1129
The Workplace Directive.
The Factory Act 1961.
The Office, Shops and Railway Premises Act 1963.
The Theatre Act 1968.
The Private Place of Entertainment (Licensing) Act 1967.
The Guide to the Fire Precautions Act No. 1. Hotels and Boarding Houses.
Electricity at Work Regulations 1989.
Fire Safety at Work – Home Office/The Scottish Office.
The Workplace (Health, Safety and Welfare) Regulations, SI 1992.
Workplace Health, Safety and Welfare, Approved Code of Practice.
Management of Health and Safety at Work – Approved Code of Practice.
Fire Precautions in the Workplace. Information for employees.

10.3 Lighting fundamentals

Light travels in a sinusoidal manner in waves. The relationship between the wavelength, frequency and velocity of propagation is:

$$\underset{\text{(metres/sec)}}{\text{Velocity}} = \underset{\text{(hertz)}}{\text{frequency}} \times \underset{\text{(metres)}}{\text{wavelength}}$$

Radiant flux is the term used to denote the total power of elctromagnetic radiation emitted or received. It is measured in watts. It can have both visible and invisible elements with the visible part being referred to as luminous.

The radiant efficiency is given as:

$$\text{Radiant efficiency} = \frac{\text{radiant flux emitted}}{\text{power consumed}} \times 100\%$$

Radiant flux which contains those wavelengths which can be seen by the human eye, is considered as having a corresponding value of luminous flux termed the lumen. It is measured by a lightmeter with a photoelectric cell. This measurement of luminous flux is called photometry.

Luminous intensity is a measure of the luminous flux per steradian in a stated direction and measured in candela (cd). Often the candela is referred to as the luminous intensity.

Illuminance (E) is the term given to the quantity of luminous flux falling on the unit area of a surface and is measured in lux. It is equivalent to lumens per square metre. The luminance (L) is the luminous intensity emitted by a specified light source per unit area and measured in candela per square metre.

Luminosity is often called apparent brightness to describe the sensation of an observer subjected to the stimulus of luminance.

10.4 Classification of luminaires

Product requirements

BS EN 60598-2-22: 1999 invokes the essential requirements.

(1) *Earthing*. All exposed metal parts are to be earthed unless the equipment is double insulated.
(2) *Flash test*. A 100% manual test regime is to be applied for 1500 volts between line and neutral for 3 seconds.
(3) *Clearances*. There must be adequate creepage and clearance distances between live parts of different polarity and between live parts and accessible metal parts.
(4) *Instructions*. Installation instructions and information for replacing serviceable parts is to be provided.
(5) *Fire retardancy*. External parts must be fire retardant and be capable of satisfying the 850°C hot wire test.
(6) *F mark*. Luminaires are to be capable of mounting on a flammable surface and be marked as such.
(7) *Light output*. The rated lumen output taking account of all correction and ageing factors.
(8) *Photometric performance*. The spacing details for luminaires calculated at various heights.
(9) *Response time*. This must be within 5 seconds of the mains disconnection and provide a light output of 50% of the rated output. The light output is to be 100% of the rated output within 60 seconds. For high risk task areas the response time is to be 100% of the rated output and within 0.5 seconds.
(10) *Brown-out*. The luminaire is to change over from the normal to the emergency operation within a band of 60%–85% of the rated supply voltage.
(11) *Battery life*. This must be designed for a 4 year minimum life span for self-contained luminaires and when operating normally and if a maintained lamp is on at 1.06 × rated supply voltage.
(12) *Marking*. Luminaires are to be marked as follows (from BS 4533):

(a) Mark of origin.
(b) *Rated voltage.* Luminaires for tungsten filament lamps need only be marked if different to 250 volts.
(c) Rated maximum ambient temperature if other than 25 °C (ta . . . C).
(d) Symbol of class 11 or class 111 luminaire as applicable.
(e) Ingress protection (IP) number as applicable.
(f) Manufacturer's model or type number.
(g) *Rated wattage.* If the lamp wattage alone is insufficient the number of lamps and type shall be given. Luminaires for tungsten filament lamps are to be marked with the maximum rated wattage and number of lamps.
(h) Symbol for luminaires with built-in ballast or transformers suitable for direct mounting on normally flammable surfaces if applicable.
(i) Information concerning special lamps if applicable.
(j) Symbols for luminaires using lamps of similar shape to 'cool beam lamps' where the use of a 'cool beam lamp' may impair safety.
(k) Terminations to be clearly marked to identify which termination should be connected to the live side of the supply where necessary for safety or to ensure satisfactory operation. Earthing terminations are to be clearly indicated.
(l) Symbols for the minimum distance from lighted objects for spotlights or such where applicable.

Figure 10.1 shows the classification of luminaires according to the material of the supporting surface for which the luminaire is designed (from BS 4533).

Description of class	Symbol used to mark luminaires
Luminaires suitable for direct mounting only on non-combustible surfaces	No symbol – but a warning notice is required
Luminaires without built-in ballast or transformers, suitable for direct mounting on normally flammable surfaces	No symbol
Luminaires with built-in ballast or transformers, suitable for direct mounting on normally flammable surfaces. This is achieved using 7 mm stand-off	▽F

Figure 10.1 *Classification of luminaires*

10.5 Battery data

Typical performance data to BS 6290 Part 4 for lead–acid cells and BS 6290 Part 2 for nickel–cadmium cells is shown in Table 10.1.

Table 10.1 *Battery data*

Nominal output volts	*Output amps*			*Output watts*		
	1 hr	*2 hr*	*3 hr*	*1 hr*	*2 hr*	*3 hr*
Gas recombination sealed lead–acid[1]						
24	12.9	7.5	5.4	350	180	130
24	19.3	11.3	8.2	463	271	197
24	23.6	14.2	10.4	566	341	250
24	31.4	19.0	13.8	754	456	331
24	39.3	23.7	17.3	943	569	415
24	49.1	29.6	21.6	1178	710	518
24	66.0	39.9	28.6	1584	958	686
24	77.8	44.8	32.3	1867	1063	715
24	97.2	56.0	40.4	2333	1344	970
24	129.0	71.5	52.5	3096	1716	1260
24	144.0	80.4	59.1	3456	1930	1418
48	12.9	7.5	5.4	620	360	260
48	19.3	11.3	8.2	925	542	394
48	23.6	14.2	10.4	1133	682	499
48	31.4	19.0	13.8	1507	912	622
48	39.3	23.7	17.3	1886	1138	830
48	49.1	29.6	21.6	2357	1421	1037
48	66.0	39.9	28.6	3168	1915	1373
48	77.8	44.8	32.3	3734	2150	1550
48	97.2	56.0	40.4	4666	2688	1939
48	129.0	71.5	52.5	6192	3432	2520
48	144.0	80.4	59.1	6912	3859	2837
108	12.9	7.5	5.4	1393	810	583
108	19.3	11.3	8.2	2084	1220	886
108	23.6	14.2	10.4	2548	1534	1123
108	31.4	19.0	13.8	3391	2052	1490
108	39.3	23.7	17.3	4244	2560	1868
108	49.1	29.6	21.6	5303	3197	2333
108	66.0	39.9	28.6	7128	4309	3089
108	77.5	44.8	32.3	8558	4928	3553
108	97.2	56.0	40.4	10692	6160	4444
108	129.0	71.5	52.5	14190	7865	5775
108	144.0	80.4	59.1	15552	8683	6383
Nickel–cadmium plastic cased[2]						
24	7.5	4.9	3.5	180	118	84
24	13.0	8.4	6.0	312	202	144
24	19.8	12.9	9.2	475	310	221
24	26.7	17.3	12.4	641	415	298
24	33.5	21.8	15.5	804	523	372

Table 10.1 *(continued)*

Nominal output volts	*Output amps*			*Output watts*		
	1 hr	*2 hr*	*3 hr*	*1 hr*	*2 hr*	*3 hr*
24	41.0	26.6	19.0	984	638	456
24	47.2	30.6	21.9	1133	734	526
24	49.9	33.7	24.1	1198	809	578
24	58.7	39.5	28.2	1409	948	677
24	68.1	45.3	32.3	1634	1087	775
50	7.5	4.9	3.5	375	245	175
50	13.0	8.4	6.0	650	420	300
50	19.8	12.9	9.2	990	645	460
50	26.7	17.3	12.4	1335	865	620
50	33.5	21.8	15.5	1675	1090	775
50	41.0	26.6	19.0	2050	1330	950
50	47.2	30.6	21.9	2360	1530	1095
50	49.9	33.7	24.1	2495	1685	1205
50	58.7	39.5	28.2	2935	1975	1410
50	68.1	45.3	32.3	3405	2265	1615
110	7.5	4.9	3.5	825	539	385
110	13.0	8.4	6.0	1430	924	660
110	19.8	12.9	9.2	2178	1419	1012
110	26.7	17.3	12.4	2937	1903	1364
110	33.5	21.8	15.5	3685	2398	1705
110	41.0	26.6	19.0	4510	2926	2090
110	47.2	30.6	21.9	5192	3366	2409
110	49.9	33.7	24.1	5489	3707	2651
110	58.7	39.5	28.2	6457	4345	3102
110	68.1	45.3	32.3	7491	7491	3553
Tubular plate mono block batteries[3]						
24	14.7	9.6	7.0	350	230	168
24	30.0	19.2	14.0	720	461	336
24	45.0	28.8	21.0	1080	691	504
24	60.0	38.4	28.0	1440	921	672
24	75.0	48.0	35.0	1800	1152	840
24	90.0	57.6	42.0	2160	1382	1008
48	14.7	9.6	7.0	700	461	336
48	30.0	19.2	14.0	1440	922	672
48	45.0	28.8	21.0	2160	1382	1008
48	60.0	38.4	28.0	2880	1843	1344
48	75.0	48.0	35.0	3600	2304	1680
48	90.0	57.6	42.0	4320	2765	2016
108	15.0	9.6	7.0	1620	1037	756
108	30.0	19.2	14.0	3240	2074	1512
108	45.0	28.8	21.0	4860	3110	2268
108	60.0	38.4	28.0	6480	4147	3024

108	75.0	48.0	35.0	8100	5184	3780
108	90.0	57.6	42.0	9720	6221	4536
Lead–acid planté[4]						
24	11.7	7.1	5.0	281	170	120
24	23.5	14.3	10.1	556	343	242
24	35.3	21.4	15.2	847	514	365
24	47.8	31.8	21.8	1147	763	523
24	65.6	43.2	28.0	1574	1037	672
24	82.0	50.0	35.0	1968	1200	840
24	98.4	60.5	42.0	2362	1452	1008
24	131.2	83.2	61.9	3149	1996	1486
50	11.7	7.1	5.0	585	335	250
50	23.5	14.3	10.1	1175	715	505
50	35.3	21.4	15.2	1765	1070	760
50	47.8	31.8	21.8	2390	1590	1090
50	65.6	43.2	28.0	3280	2160	1400
50	82.0	50.0	35.0	4100	2500	1750
50	98.4	60.5	42.0	4920	3025	2100
50	131.2	83.2	61.9	6560	4158	3095
110	11.7	7.1	5.0	1287	81	550
110	23.5	14.3	10.1	2585	1573	1111
110	35.3	21.4	15.2	3883	2354	1672
110	47.8	31.8	21.8	5258	3498	2398
110	65.6	43.2	28.0	216	4752	3080
110	82.0	50.0	35.0	9020	5500	3850
110	98.4	60.5	42.0	10824	6655	4620
110	131.2	83.2	61.9	14432	9147	6809
5 Year life sealed lead–acid						
24	10.0	8.5	6.0	240	204	144
24	23.0	14.0	9.0	552	336	216
24	30.0	16.7	12.0	720	400	288
24	40.0	23.0	18.0	960	550	432
24	80.0	46.0	36.0	1920	1100	864
48	10.0	8.5	6.0	480	408	288
48	23.0	14.0	9.0	943	672	432
48	30.0	16.7	12.0	1002	801	576
48	40.0	23.0	18.0	1271	1104	864
48	80.0	46.0	36.0	3840	2208	1728
108	10.0	8.5	6.0	1080	918	648
108	23.0	14.0	9.0	2484	1512	972
108	30.0	16.7	12.0	3240	1803	1296
108	40.0	23.0	18.0	4320	2484	1944
108	80.0	46.0	36.0	8640	4968	3888

[1]Sized to 1.7 v.p.c. at 20°C Ambient Temperature Complying with BS 6290 Part 4.
[2]Sized to 1.02 v.p.c. at 20°C Ambient Temperature Battery life up to 25 years.
[3]Data at 25°C to 1.7 v.p.c. 85% of nominal. Battery life up to 10 years.
[4]Sized at 1.7 v.p.c 20°C (as per BS 5266) Battery housed on double tier double row steel rack.

Part 2 Security Lighting Systems

11 Architecture

It is not possible to clearly define security lighting in the same way that we have established the role of emergency lighting, and we have now covered in some detail. In addition we are not able to produce standards that will govern how it is to be specified and installed. In practice there are a number of reasons for this. Security lighting will be found to exist in many diverse forms and it is not possible to quote particular levels of illumination that are needed because these vary across the different applications. Indeed these applications may be as simple as the lighting of a small external domestic area but can be as complex as providing high levels of illumination to support CCTV observation systems in highly sensitive and high risk environments. In practice any lighting provides an element of security lighting even though it may be quite marginal. Statistics and surveys prove that intruders are intimidated by light because it always makes them visible even to the casual observer who may then report the incident. From this we can say that the object of security lighting is to:

- Improve the levels of illumination and heighten the safety of authorized personnel.
- Support the efficiency of other security and safety installations.
- Improve the likelihood of detecting unauthorized personnel and intruders.
- Deter crime by providing illumination of an area.
- Provide a welcoming environment for authorized users.

As an overview it is possible therefore to state that security lighting is intended to illuminate areas in order to deter crime but also to provide a safe and reassuring environment. In many applications it may also include amenity lighting so that the security lighting can be used to provide lighting for general illumination duties.

It can be observed that there are no exact standards to refer to when specifying security lighting but the components and the requirements of the installation will be governed by the general standards in force within the electrical industry.

Although it is possible to introduce an element of security in any application by manually introducing luminaires, which are effectively the complete lighting unit, via normal switches, this is not generally the case because we must then rely on human activity to select the desired functions. For reasons such as this lighting for security and safety purposes tends to be selected automatically but normal manual switching is often added to override the automatic function for the amenity role.

Lighting as a security technique also has a role to play in many management systems in which we have a bias towards energy saving. These systems are increasing in significance and are covered in Chapter 18.

Security lighting is often linked to other technologies and time honoured security techniques to form part of an integrated system and the means of doing this are detailed in Chapter 19. By amalgamating lighting with other functions it can be used to even greater effect and its use then becomes even more diverse. Lighting when used with a series of other components or systems can further assist in protecting both personnel and property.

Although security lighting does have an extremely wide use and wide ranging function in the main we have two broad categories. These are namely demand lighting and extended period lighting.

11.1 Demand lighting

This is lighting that under normal circumstances is activated by an automatic detection device that will keep the lights energized for only a short period of time. Therefore this type of lighting is a form of pulsed lighting and is brought into effect for only a prescribed amount of time that is determined by a timer circuit. This timer may be incorporated in the detector itself or it may be at a remote location in a purpose-designed controller.

The most popular detection technique of signalling demand lighting is using passive infrared (PIR) sensors. These sensors may be integrated with the lamp or they may be mounted at a separate location. It is not always possible to obtain the optimum position for both the lamp and the detector using integrated assemblies and it is often the case that the two components need to be mounted at separate locations. This technique allows the installer to select the ideal position for the different components. Integrated or self-contained PIR switched lights make up a high percentage of the budget domestic security lighting industry because they are available in a multitude of forms. These extend from decorative carriage lights and bulkheads through to tungsten halogen floodlights. Recent additions to the range even include cameras that form part of the assembly and enable connection to be made to existing VCR equipment for automatic recording and verification of an event taking place. The sensors themselves feature options on a number of detection patterns. The assemblies also include photocells so that the lux level at which the luminaire will be activated when the sensor is tripped can be set as appropriate to the site conditions.

These simple standalone integrated assemblies are intended to illuminate an area when a person or vehicle is within the coverage pattern of the sensor. The lights will remain energized for a prescribed period of time when motion within that area has ceased. More advanced systems use separate sensors that may be mains powdered or an option is to use extra low voltage sensors that have a unique controller. These specialized controllers allow a number of sensors to be connected and enable individually managed zones to be achieved. They also allow extension warning buzzers to be attached to give audible tones that a sensor has been activated. A further advantage of using extra low voltage sensors is that the wiring is signal cabling so it is not governed in the same way as cabling that is used for mains powered devices.

Although the PIR is the most widely adopted detection device for demand lighting there are a number of alternative sensors that can perform the same role and these are considered in detail in Chapter 14. Indeed any automatic detector or sensor can be used to energize demand lighting. These alternative sensors may come complete with power supplies and heavy-duty relays to link with lighting systems. Otherwise they may form a part of a greater network and additionally bring on lighting for a specific area when intrusion in that locality is detected. Intruder alarm systems often switch on demand lighting when they are alarmed so that an area is illuminated. They may also be linked to a CCTV monitoring system so that a specific camera is invoked for that area.

Although the most widely used sensor technique, PIR detection is not necessarily the best option for every application because it does have considerations of installation in order to ensure that sporadic and unwanted signals do not occur. If the sensitivity or stability of the passive infrared can create a problem or if the range of the PIR is unsatisfactory an alternative sensor should be used. To protect fences, barriers and long boundaries it is a better option to use more specialized detection techniques or active infrared beams. All sensors can be linked to lighting systems so it remains to say that it is essential to select the most appropriate technique otherwise the demand lighting is rendered useless.

Although there is an overlap of some luminaires used for both demand lighting and extended period lighting, there are other lamps that are not suitable for both types of duty. As an example, low energy lights or those with ignitor ballast circuits are not intended for demand lighting but are for use over extended periods. These low energy lamps are not intended to be switched on for short periods because of their starting characteristics.

There are certain factors to be considered before specifying demand lighting:

- Demand lighting can have a shock effect to both users and intruders.
- It has no early reassuring effect.
- It can cause glare and startle personnel if incorrect lighting is used.
- Sensors should be sited to energize the lighting well in advance of the sensitive areas it actually protects.

There will always be a healthy installation market for demand lighting and it will continue to be cost effective and energy saving because lighting is only made available for short periods of time. In addition lights are not left on at unwanted periods. Demand lighting should also have a facility to be overridden so that the lighting can be manually switched on for special events and occasions.

11.2 Extended period lighting

This is for areas that require lighting to be available for long periods of time such as from dusk to dawn. Control can be by manual switches, separate or integral photocells or by timers. Although in practice any lighting type can perform this duty for reasons of economics it is best served by low energy luminaires.

The great advantage of extended period lighting is that it provides a reassuring effect because lighting is always available and there is no startle generated by lighting suddenly becoming available when personnel enter a specific area. However, there is a need to select the correct lighting type if energy savings are to be made and maintenance costs are to be acceptable.

Light pollution must be taken into account when specifying extended period lighting because of the long running times in which it is employed. A serious problem is now being caused by light pollution and it is considered by many observers to be as offensive as noise, smoke and chemical discharge. It is generally the result of badly designed lighting systems and inaccurately sited and aimed luminaires. It causes the lighting to spill over and trespass into areas in which it is not intended to be, including the night sky. A consideration of the specification is therefore selection of the correct light colour appearance and ways in which stray light can be avoided. If CCTV systems are to make use of the lighting pollution could be a problem and when colour rendering and discrimination is important it is necessary to match the cameras to the lighting. Infrared lighting has advantages in negating pollution and can be considered for monochrome cameras. This is covered in detail in Chapter 17.

For high security applications and where light pollution cannot be a problem disability glare can be achieved with extended period lighting. Large high intensity lamps at high level can cause disability glare to a criminal trying to enter a specific area. This intimidates the intruder and produces indecision because an unauthorized person is unable to determine what barriers are in place or where security personnel are operating. For authorized persons to have good visibility there is a need to ensure that the lighting does not cast shadows as these shadows themselves provide areas for intruders to remain undetected.

Having come to appreciate the main aspects of demand and extended period lighting it is necessary to balance the advantages and disadvantages of the two different types before deciding on which is best for each application. In practice there are many issues to be considered but all are dependent upon:

- The best lighting form in the area and the luminaires that can provide this.
- Types of surfaces that are to be illuminated.
- The need for CCTV monitoring and colour discrimination.
- If demand lighting is to be used the distance at which detection is to be achieved.
- The capability of detectors in providing effective demand lighting.
- Benefits of permanent illumination in the absence of natural light.
- Light pollution and the effect of it on adjacent buildings and neighbourhoods.

The actual level of illumination that is employed for the different duties and in the various applications will be found to vary enormously. In some instances it will be formed from large floodlights illuminating wide areas with a white glare whereas in other applications the lighting shall be found to be of the minimum lux levels and perhaps also using directional beams.

External lighting is changing because light pollution spoils the effect of a natural night sky and has an effect on the environment. There are also considerations of power consumption and the use of energy. However, external lighting remains essential for safety and security applications to include illumination for CCTV and monitoring duties. In addition it has a role to play in decorative installations that may be integrated with these security roles. It is a question of balance to determine the safe environment with minimal light pollution and interaction with the purpose for which it is being specified. It is correct to say that lighting schemes have to be innovative and intelligently designed and yet provide a level of vandal resistance without creating any high level of pollution.

Light remains one of the best deterrents to crime and is increasing in importance to support the growing use of CCTV systems. We are also seeing a greater employment of low energy lighting at the expense of high brightness illumination sources and we expect this trend to continue. The surveyor when selecting any security lighting scheme is obliged also to show an awareness of energy management and occupancy lighting that can be used in a safety mode.

In all of the systems that we are to encounter there will be numerous wiring techniques employed and these will go through the ranges of extra low voltage up to systems that are mains wired with the loads cabled on different phases of mains multi-phase networks.

Security lighting offers great potential to both groups, the installers of the product and the end user. It has very much an overt role so the client perceives it as a good investment as it is seen to be in regular use particularly when adopted also as amenity lighting. The duty must always be to install systems of quality as its image can easily be tarnished by poorly installed lighting. Extra attention should be directed at the problems of high illumination levels in inappropriate circumstances.

We therefore can accept security lighting as a serious security solution and having an additional amenity role. However, we must not lose sight of the wider picture and must therefore consider all of the options before deciding on the means by which the technologies are to be introduced. The first stage is to look at the systems that are available in order to select the most suitable version and technique.

12 Selection of lighting system

The ideal of all security lighting systems is to improve vision and thereby reduce the incidence of crimes that are committed during the hours that natural light is unavailable. To the perpetrators of crime the effects of darkness conceal their activities and diminish the risk of being detected. A lack of illumination or natural light also ensures that they cannot be easily recognized or observed.

Lighting helps to negate these problems and takes away the areas of concealment. In addition it acts as an aid to safety by illuminating walkways and ensuring that obstructions can be seen. Security lighting forms a huge subject because it may be as simple and basic as one lamp illuminating a small domestic point yet it may be as complex as a technique integrated with other security and building management systems in high security external schemes. There are large markets for schemes that are between these two extremes and these of themselves can be linked to many other functions and technologies in order to make the most of opportunities presented by the other building systems. As an example it is easy to link lighting to intruder, access control and CCTV techniques so that they complement each other. This can be done by basic interfaces such as relays or by programmable outputs that are dedicated to other functions. There will also be flexibility offered by purpose designed lighting systems to enable the facilities of such schemes to invoke functions from the other networks. Sensors that are used to energize lighting schemes can often also signal outputs from other systems so that the sensor can, for instance, provide an audible signal and start up CCTV equipment by being linked with intruder and observation techniques.

In keeping with any security concept the specification for the lighting must be linked to the level of risk that exists. It is more difficult to define security lighting than many other security functions as there is no specific standard that relates exactly to the different forms. However, there are many standards that govern the wiring techniques and the products that form the system components. From this we may conclude that we have classifications which broadly relate to:

- Lighting that is intended to detect intruders and enable them to be observed. This is called demand lighting and is available for

only short periods of time. Chapter 15 covers the fundamental issues.
- Lighting that is used to deter attempts to enter an area. This is lighting that is kept energized for long periods of time and gives a reassuring environment. It is generally called extended period lighting and makes use of low energy luminaires. This is detailed in Chapter 16.
- Lighting that is intended mainly for safety in order to allow the safe passage of guards through areas and of other authorized personnel. It is similar in use to that of extended period lighting.

The level of illuminance or amount of light will vary in every system depending on the actual type of luminaires employed and the risks envisaged. This is governed by the protection that may have to be provided for the actual lights to ensure that they are adequately safeguarded from damage. This equally applies to sensors that are used in demand lighting to detect movement and provide the signal for the lighting to be energized or for control equipment used for extended period lighting systems. There are a number of factors that need to be considered for all the levels of risk and the extent to which the lighting components themselves must be guarded. These can be identified under the subject headings of:

- Risk
- Automatic or manual operation
- Type of luminaire
- Wiring forms
- Level of illuminance
- Capacity of system
- Integration
- Budget.

12.1 Risk

The system is to be related to the level of risk and probability of attack or danger. This can include the value of goods within an area or the strategic importance of the area, premises or buildings.

There are general levels that can be used to grade the security class as illustrated at Table 12.1.

There will always be an overlap of products used for the different classes of risk, and security lighting can also be found as a product of other systems. It may be brought into effect when a different system has responded such as an intruder alarm activating and then triggering the lighting network. By itself lighting cannot be adequate in every

Table 12.1 *General level of security class*

Classification	*Considerations*
Basic security and amenity	Low risk using standard luminaires to provide convenience lighting and allowing a small measure of security to be achieved.
Mid risk security	Using luminaires that have a greater resistance to vandalism and sensors or photocells/timers that are less accessible to vandals/criminals. Considerations given to protection of the mains supply and adoption of multi luminaires to compensate for lamp or component failure.
High risk	Vandal resistant luminaires and sensors or photocells/timers that vandals/criminals will have difficulty accessing. Cabling and the mains supply is to be protected. A greater number of luminares to be used to cater for component failure. Systems can be linked to outputs of other security systems to include CCTV monitoring.

application and it must often be supported with other security technologies if a high level of protection is to be afforded. We must also remember that in certain instances the security lighting function is automatic but in other cases it may be purely manual.

If the level of risk is particularly high it is possible for certain parts of the security lighting system to be supplied by a generator with an appropriate battery and inverter. If a central inverter is being used to energize emergency lighting in a given premises and if it has adequate spare capacity this could be employed to energize a small number of security lighting luminaires for critical applications. An advantage of central inverters is that they are designed for use with mains type luminaires and can provide mains power of a similar voltage and frequency by inverting the standby batteries potential to ac in the event that the normal mains supply is disconnected.

In the petrochemical and oil exploration industries there are special risks and requirements that govern all electrical equipment. These are known as hazardous areas because they may contain flammable gases, vapours, liquids or solid substances which can burn either slowly or in an explosive manner depending on the substance and prevailing conditions. These areas may at times also be called locations having a flammable atmosphere or a potentially explosive atmosphere. They extend to any industrial site that can be hazardous to include the mining industry together with paint or chemical processing plants.

The term explosion protection is applied to items of electrical equipment that are intended to be installed in hazardous areas and are so designed and constructed that they are not capable of causing ignition of the surrounding atmosphere when installed as intended. Various methods of preventing this ignition are used and they are all termed methods of explosion protection. The selection of the specific method of explosion protection is dependent on the zone classification. This is a concept of recognizing the differing degrees of probability with which explosive concentrations of flammable gas or vapour may arise in installation in terms of both the frequency of occurrence and the probable duration of existence on each occasion. The three levels of probability are classified by zones:

Zone 0 An explosive gas–air mixture is continuously present or present for long periods
Zone 1 An explosive gas–air mixture is likely to occur in normal operation
Zone 2 An explosive gas–air mixture is not likely to occur in normal operation, and if it does, it will only exist for a short time

Mains operated luminaires are not allowed to be used in zone 0 areas other than beta lamps. The beta lamp is a self-illuminating light source and does not need an electrical supply to operate. Although these lamps do not require an external energy source the light output is low as they use a special glass capsule and internal phosphor with tritium coating.

The standards that apply are BS 5345 and BS 6467. These cover the protection for those areas in which gases and dusts can be hazardous and specify the methods of explosion protection or type protection that the equipment must be certified or approved to by the relevant certification authority. The symbols are prefixed Ex to show their status.

The types of protection in which we are interested are listed in Table 12.2.

Legislation places the responsibility for complying with safety requirements on both the manufacturer and the user so the normal method of demonstrating compliance is to select equipment that has been tested and approved by a recognized authority for the area in which it is to be used.

In the UK the British Approvals Service for Electrical Equipment in Flammable Atmospheres (BASEEFA) is the recognized approval authority and grants certification to the appropriate standards and classifications. They offer 3 yearly renewable Certificates of Assurance.

All luminaires that are granted approval for use in hazardous locations are allocated an ignition temperature classification according to

Table 12.2 *Hazardous areas type protection*

Type	*Description type of protection*	*Use*
Ex d	Flameproof enclosure	Zone 1
Ex p	Pressurized enclosure	Zone 1 and zone 2
Ex e	Increased safety	Zone 1 and zone 2
Ex s	Special protection	Allows development of new techniques
Ex N	Non-sparking	Zone 2

its surface temperature under given working conditions. The temperature class of the luminaire is not to exceed the ignition temperature of the gas. This is because the heated surface of the luminaire could cause an ignition of the gas. The user must always accept a measure of responsibility in selecting a luminaire for working in any particular environment to take all of the factors governing the approval into account.

The temperature class and its relationship to the maximum surface temperature is as shown in Table 12.3.

The certification of the electrical apparatus would also cover the equipment for use with certain gases and vapours and these would be divided into subgroups with representative gases. The gas group would be allocated and marked on the product in the same way as the zone, type of protection and temperature class.

Luminaires certified for use in hazardous locations would therefore be of high quality heavy-duty form and capable of working in extreme conditions. For luminaires to carry out the role of security lighting in industrial areas such as marine oil and gas terminals, chemical, petroleum and gas processing plants certified units must be used and these may have conversion packs if intended to work as emergency lamps.

Table 12.3 *Relationship of temperature class and maximum surface temperature*

Temperature class	*Maximum surface temperature*
T1	450°C
T2	300°C
T3	200°C
T4	135°C
T5	100°C
T6	85°C

12.2 Automatic or manual operation

Although certain systems are purely manual in their switching technique it is normal to use an automatic mode to some extent. By classifying systems as manual in their operation it suggests a means of using switches to turn on the luminaires at a given point in time. Although we tend to have manual switches in security lighting these switches are generally used to override the automatic function only so that lighting is available for the time duration selected by the switches. However, it is more normal that the lighting is selected automatically with the manual switches as an option. This means that the system does not have to rely on human input to select the lighting. The manual switches ensure that whatever is the state of the automatic function the lights may still be activated for specific conditions. In the respective Chapters 15 and 16 dealing with demand and extended period lighting, details are included of how the manual switching is applied to complement the automatic role.

The automatic means of selecting lighting is by timers and photocells for lighting that are to stay energized for long periods of time. For lighting that is only intended to be energized for short time scales this is achieved by sensors that are considered in detail in Chapter 14 which relates to detection techniques.

A considerable range of timers is available to select security lighting. These may be electronic in operation and allow settings and modes to be made that can be random in their working pattern. By this means the sequence of the times that the lights are energized change on a regular basis so that a potential intruder can never be certain as to whether a premises or area is actually inhabited by security personnel or owners. Such electronic timers use programmable techniques and the settings can be displayed on an integral screen or at a remote point. The timer is used to signal contactors or relays which then drive the lighting through appropriately rated contacts.

Electromechanical timers tend to be set using tappets. They have a time clock to show the times that have been set in a visual sense. Timers often need to be updated throughout the year to compensate for changes in light levels in the seasons.

All timers developed for the commercial and industrial sectors have respectable contact ratings generally in the order of 16 A at 240 Vac. These often use single throw contacts but they may have changeover switching elements for identification purposes or other functions.

Photocells are rather different to timers as they do not need to be adjusted as the year moves through its seasons. This form of automatic control is achieved by a device that has a light sensing photoelectric cell. These detect levels of prevailing illuminance on the working plane and using a form of sequence switching they turn off the contacts when the illuminance on the working plane exceeds a preset value. This value is often adjustable to suit the location of the device. In addition fail safe devices can ensure that if the photocell develops a fault the lamps will remain on to show a system malfunction.

These photocells are available in a range of sizes and with options on lux settings (generally 2–10 lux). They will have different switching capacities and levels of protection to the weather. In addition they compensate for minimal changes of ambient light so that for instance car headlights do not falsely trigger the unit. They may also be referred to as twilight switches.

In addition to the time-honoured functions of time and photocell settings some products specify brightness control. These combine the techniques of photocells and sensor detection. This brightness control suggests an individually adjustable basic brightness of 1–100%. It means that as an example a device can illuminate an entrance area throughout the night with low wattage lighting but when approached by a person or vehicle can then switch to a higher wattage. When the latter leave the detection zone the device will revert back to its original brightness. If in contrast the device is set to 0% the luminaires will only respond to movement in the detection zone.

12.3 Type of luminaire

This must be balanced to the level of risk. There are a number of important considerations to be made to include aspects such as bulb downtime and the use of a greater number of luminaires to account for component and unit failure. In low risk situations and in those circumstances where only a small measure of security is involved as the lighting is mainly for amenity purposes standard luminaires can be used. In these situations aesthetics may also play a part. However, as the level of risk heightens the surveyor must take into account such factors as luminance, contrast and glare in addition to impact resistance of the fitting and its siting in a normally inaccessible position. In each case and according to the laws on the safety of equipment, luminaires are technical products and must be used for their intended purpose only. The actual intended purpose of each variant will be declared by the manufacturer of the goods.

We can regard illuminance as the density of the light being projected into a given sector and this forms the basis of the design. Due regard must also be given to fog and other environmental problems that can effect the lighting of that area at particular points in time. There will be a need to ensure that uniformity of illumination is achieved over the area being protected. This is better achieved by using wide luminance pattern lamps in preference to smaller units with narrower projection patterns. The appearance of the person in the background of the area is effectively a combination of the contrast between the actual illuminance of the person and the background. This contrast is that of the level between the shades of white, through grey to black. Good contrast is needed to give definition and to enhance the colour and differential.

Glare or excessive brightness depends on the contrast conditions and the direction that the observation takes. For this reason the lighting must be directed into the field of view to be observed and it must not cause a light pollution by being also guided into the level of vision of other unrelated areas or premises.

All luminaires have some degree of impact resistance but this becomes of increased importance as the level of security becomes greater and as the risk of determined attack heightens. In high security applications it is more apparent that the luminaires need to be positioned at higher levels so that it is more difficult to gain access to them without the use of additional equipment. The security of the luminaires can be further enhanced by employing restrictions beneath them that would need to be removed before platforms/ladders or such can be placed at the base of towers on which floodlight projectors have been installed.

In those cases in which it is not possible to install the luminaires at high levels or in positions that are difficult to get access to, the lamps must be of a special vandal resistant construction. There are now ranges of luminaires that can be flush mounted in walls by the use of recessing boxes. These lamps are held behind vandal resistant louvres. The louvres come complete with toughened glass and are high impact. By use of this technique the lamps and control gear are held some distance back from the actual louvre so that any impact is not directed on the lamp or control gear. In addition the lamp and the associated control gear and wiring is not heavily vibrated by the impact of attacks on the louvre or the surrounding building structure.

These vandal resistant luminaires come in a variety of lamp options and have clearly defined cutouts in the material into which the recessed housing box is to be held. These luminaires also provide high levels of protection to the ingress of dusts and moisture and give good levels of security and safety orientation even though the luminaire may only be mounted at a height that can be accessed without the need for ladders or climbing aids.

With any lighting system there is a need to select luminaires that are both safe and of high quality and are manufactured to comply with the relevant standards. Unfortunately there is always a significant amount of substandard equipment being sold in the market place and the implications of using this equipment are plain to see. There must be a drive to install only the best quality components to ensure that the best performance is achieved for the customer.

In keeping with selecting high quality luminaires there is a need to make sure that the lighting will:

- Illuminate the exact area to the required level.
- Be located at the most appropriate positions.
- Provide security and not develop any dangerous situation.
- Be of the correct classification.

In dealing with the latter category of luminaire classification it must provide the correct type of protection against electric shock plus an appropriate degree of protection against the ingress of dust or moisture. It must also be rated to the characteristics of the material of the surface to which it is to be secured. Therefore in theory and practice the luminaire must be able to operate safely and to withstand the environmental conditions in which it is to carry out its desired function.

The vast range of duties that are covered by security lighting installations leads to a use of lighting types across the full range of available luminaires and these are covered in Chapter 13.

12.4 Wiring forms

Unlike emergency lighting there can be no exact standard to invoke the type of cable or the protection it must afford. Although there is no specific practice that we are able to refer to we can say that the cabling is to be specified to take account of size, rating, routing and protection. It may in some instances be extra low voltage cabling employed for what we term as low voltage sensors and in other instances it may be tamper protected to monitor for cutting or bridging.

In many instances standard PVC sheathed and insulated cables will suffice but in other applications it may be appropriate to use armoured cable or to enclose the wiring in steel conduit. This depends on the likelihood of it being subject to abuse or to combat environmental challenges. In other respects the cable is governed by the normal duties of electrical engineering practices. It is always ideal if the cabling is not accessible under normal circumstances as this ensures its integrity and security. In addition there is no need to add containments such as conduit tubing or trunking so the installation remains equally high in its aesthetic value.

The cable is to be rated to compensate for the loads of the lamps and some consideration should be given to the number of circuits that control the loads and how these may be spurred in parallel. It is always necessary to have dual cabling used for loads in high risk applications so that the lighting fittings do not rely on a single supply cable. In low risk domestic applications when cabling is difficult it is often the case that the sensors are of a low voltage form so that mains rated cables only need to be used for the wiring of the luminaires. The sensors can be cabled in a more easily routed extra low voltage cable type not governed by the more onerous requirements of those applied to mains cables devices.

The mains wiring is therefore governed in the same way as the general cabling for the building services together with the circuit protection and earthing. There may also be a need to encase wires in conduit or in protective trunking to afford additional protection or for purposes of aesthetics in areas where it would otherwise be visible or exposed to attack.

From this it can be added that maximum cable concealment is advisable and options are to use protected cable such as MICC cable or to use conduits. These methods all minimize the possibility of the supplies being cut. All installations are to comply with the IEE Wiring Regulations and be of high quality with high integrity wiring. The fitting of ring or duplicate circuits provides a further measure and path in the event of installation breakdowns either accidental or deliberate. Other measures of protecting the wiring can be by monitoring using additional supply cores and connecting these to contactors or continuously rated relays to check for loop integrity. These can bring in warning devices if the loop becomes detached.

12.5 Level of illuminance

It is essential to select the correct level of illuminance for each and every application to enhance the observers' performance and to allow objects to be properly seen. There is a need also for colours and forms to be distinguished. The human body can only detect an object based on the luminance of the subject and the surroundings or between the different points of the same object. Errors will always be attributed to individuals or the security lighting will be judged inaccurate as a consequence of illuminance levels being too low. We must strive to achieve the optimum value of illuminance and quality of illumination.

In selecting the level of illuminance we can confirm that the principal aim of any security lighting is to:

- Heighten the probability of detecting and identifying unauthorized persons whilst improving the security and safety of authorized persons.
- To improve the efficiency of other measures, building management services and security systems in use.

In general the lamps for exterior lighting must be positioned so that they do not cast any shadows which may serve as a cover for a potential intruder. We can also say that all protective lighting that contrasts to complete darkness has a certain deterrent effect but this must always be arranged in order that this illumination cannot possibly be controlled by unauthorized personnel. By providing this around a building perimeter a sound line of defence is achieved. Such an arrangement is sufficient warning to an intruder who must pass through a zone of light to gain entry. Back entrances and exits are all too often overlooked as being unimportant but it is often at such points that stolen goods are removed from premises that have been burgled. When protecting the outside of buildings whether external lighting should be directed towards or away from the building has merits in both ways. If the luminaires are directed towards the building this means that direct entry is clearly seen and any lit figures stand out by the sweep of light. If the light is directed away from the premises the intruder is exposed to the glare of the luminaires in contrast to guards who would be in an unlit position.

Large open spaces always need good levels of light. Areas of this nature include major plants, military establishments, aerodromes, goods yards, oil terminals, chemical works plus car and lorry parks. The exact site conditions will dictate the best floodlighting to be employed and the need for using CCTV or if this could be envisaged for the future.

For maximum effect the level of illuminance must be determined for all of the areas and this should not readily be changed. There is a wide choice of luminaires that can act as security lighting components. Some give a simple light directed in a forward plane while others may also have an additional light towards the side.

In systems in which the risk of security breaches is low the actual level of illuminance and the quality of illumination is not as important as those in which it is necessary to identify perpetrators of crime. In every case an adequate level of illumination and quality is needed but in observation systems this takes on extra significance.

During the hours of darkness CCTV cameras for observation systems must be capable to produce clear and usable recordings of images as they occur and to this end artificial lighting has to be employed. The system is committed to generating video recordings with good resolution and contrast. This can be achieved by using a variety of different lamps that range from fluorescent through to infrared. The extra significance of this is covered in Chapter 17.

12.6 Capacity of system

All systems are restricted in the capacity of lights that they can control and in the number of sensors that can be used. The cables that are used have a maximum working capacity and there is a need also to employ appropriate circuit protection. It is advisable to include facilities in the cabling network and switches to enable parts of the larger system to be turned off or isolated for those periods when there is no need for the full system to operate. This also makes maintenance and fault finding easier.

In those systems where demand lighting is used it is inconvenient to interface a multitude of mains sensors with a correspondingly high number of luminaires. It is advisable to make up a system that incorporates zones or wards as an option to bringing on a high number of perimeter lights some of which may not be in the vicinity of the location of the sensor that has been activated. If it is important to use a high number of sensors then it is important that control-indicating equipment should be added to signify the sensor that has actually called up the lighting.

If a low voltage sensor network is being used the controller tends to have an analyser to signify the actual sensor that has been tripped. This is important if sporadic operation is occurring because there will be a need to establish at exactly what point the fault lies.

For those systems that have warning devices such as buzzers that operate when the sensors trip through the hours of natural light, these should be programmed so that the luminaires only come on through the hours when natural light is not available. If there is no special controller available to achieve this it can otherwise be easily accomplished by taking the luminaire circuit through a photocell so that this device isolates the main load until light levels are low. Balanced to the logic of ensuring that the capacity of the system is satisfactory is a need to look at energy saving and management. For this reason only low energy lights, which are intended for use over extended periods, are to be used for full dusk to dawn operation. Incandescent lights are not economical for this purpose.

In multi-phase systems the luminaires can have their security improved by using alternate phases and there is an extra need to balance the loads. When discharge lamps are used there are additional considerations with regard to current surges on start. This also has an effect on switches, fuses and circuit breakers and these must be rated accordingly.

12.7 Integration

This is the art of combining diverse elements into one whole or collective central control.

It may be the joining of different systems to allow the sharing of information and can be seen as a system that covers all possible security as well as building control applications. It can be as simple as connecting an alarm panel to an access control system, lighting and CCTV system to stop and identify persons entering that area when it is armed. It may be accomplished by using pairs of wires from the alarm panel output to the other controllers that secure access for that area. This is achieved by the use of relays. In other cases it can be as complex as having a single point of administration for all of the systems in a building. In this latter case software is created so that the user interface reflects all systems that exist and how they have been installed.

The reasons for integration are:

- There is a single administration point.
- One network.
- Tailored systems can be achieved.

Lighting has a major role to play in being linked to CCTV cameras but it is also often used in conjunction with intruder techniques. In these applications if an alarm is triggered the external lighting can also be energized to illuminate an area that an intruder may attempt to use as an exit route. In other ways the lighting can be used to signal an alarm panel when a sensor brings on the lighting. It must be noted that with external sensors this activation should be part of a dual technique to ensure that environmental disturbances and animals cannot cause false alarms.

Lighting may also be used to follow the route through an access control system if attempts to defeat the perimeter are made. In certain systems designated outputs are provided to link the different systems to each other but if these are not available the voltage-free contacts of a relay coil can be employed. The coil voltages of relays are available to suit all of the recognized supplies.

The linking of lighting to other security technologies and to other building management facilities always makes even greater use of the security luminaires. It also justifies further the cost and budget for the installation of the security lighting.

12.8 Budget

Every lighting system is bound by financial constraints so there is always a need to perform a cost analysis. It is also normal to plan the scheme so that the lights are used for amenity lighting in addition to their security function. This ensures that they are envisaged as being used to a greater extent than they would be if employed for security purposes alone.

The cost analysis should nevertheless consist of:

- Capital expenditure on equipment and labour.
- Maintenance. This to include lamp replacement and cleaning of luminaires.
- Running costs.
- Reduction in insurance premiums.

In relation to many other technological systems security lighting is not complex and is easily understood. In addition these systems are inherently reliable. The most complex items that form the system are the sensors. The luminaires are robust and time honoured in their construction and timers and photocells are reliable. We can also say that switches are not capable of causing any particular problems and this also applies to cabling if correctly installed and protected. For these reasons the ongoing maintenance and repair/replacement costs are relatively low for the advantages offered by the security lighting network. The only problem can be bulb downtime although this can be difficult to judge. Maintenance is essentially a management responsibility although employees are responsible for reporting faulty equipment that may cause danger. All electrical equipment throughout time must deteriorate to some extent so a drop in the illuminance properties must occur unless corrective action as part of a maintenance plan is undertaken. With security lighting this need not be expensive. The replacement of worn lamps increases light output and safety is improved by the removal of faulty wiring and fittings.

By the instigating of a maintenance programme it is possible to introduce changes and system upgrades:

- More modern luminaires can replace old fittings. Aesthetics and perception are heightened.
- Improved illumination increases the level of security and safety.
- An opportunity arises to replace traditional luminaires with low energy lamps.

When security lighting forms part of an integrated system it should be budgeted for separately and it should also be installed so that it can be controlled outside of the core system in the event that faults occur.

There is no expectation of significant advances being made in this technology in the foreseeable future so changes are not expected to be radical. This aids the budgeting process as the systems now being specified will have a high life span. If rapid advances were to be made in sensor and luminaire technologies this could have no effect on the cable installation which would still be employed to support any new generation of components.

When considering the budgeting we can add that energy management can also be achieved by the use of extended period lighting that gives reduced maintenance. This forms the subject of Chapter 18.

13 Luminaires and light forms

Although in practice any light form can be used to provide the security function for every application there is an optimum type. In addition there are what can be described as standard forms that are installed on a routine basis. Therefore we become familiar with certain versions that have survived the test of time and become time honoured. For observation purposes, when CCTV is in operation, the selection and performance of the lighting is even more critical. In normal daylight, activity can be easily captured on camera, but during the hours of darkness artificial lighting must be applied to achieve pictures with good resolution and contrast. This can be achieved using various lamps. These range from infrared through to fluorescent.

It is important to note that both incandescent and discharge lamps can be used in security situations.

Incandescent lamps are those that rely on the passage of an electrical current through a metallic filament for their operation. This passage of current causes the filament to become heated and to emit radiation. The range of incandescent lamps includes standard filament and tungsten halogen lamps.

Discharge lamps, however, fall into a different category as they rely on the transmission of an electrical current through a gas or vapour for their operation.

Lighting in many non-CCTV-based systems can be decorative and of display form for amenity purposes and then double up for the security function. This is subject to the specific application and capacity of the system.

The first stage is to consider the most popular luminaires and light forms in order to select the most appropriate for the particular duty. There is only a need to investigate those luminaires that are used extensively because we can reiterate that any lamp can give some measure of security no matter how small this may be.

The lighting that we are about to consider is an overview because manufacturers' specifications and technologies do differ to some extent but the essential data is similar. In every case the wattage and type of lamp will be determined by the manufacturer of the luminaire and this must be adhered to. Incorrect lamps may cause inefficiency, change the planned distribution of light, create flare or even seriously overheat the fitting.

Light forms are classified broadly as those for essential duties or general purposes and those that are known as discharge lamps.

When an overview of the light forms has been undertaken we may then consider the classification and some of the other details that affect the selection of the specific luminaire for any particular role.

13.1 Essential duty lamps

Tungsten (GLS) bulbs

These are incandescent lamps that produce light as a result of the heating effect of the electrical current flowing through a filament wire. The lamp filament when heated to a high temperature becomes white hot and emits radiation through the visible spectrum with a bias towards a high wavelength. The most popular version is the coiled coil tungsten filament bulb referred to as a general lighting service (GLS) bulb. This produces both visible light and infrared energy when the temperature is high enough. Tungsten is used as the filament because of its high melting point and low rate of evaporation at high temperatures. The GLS bulb is used extensively in the market place and is a familiar, standard household item.

The 60 mm pear-shaped bulb is time honoured as the standard GLS lamp but other shapes are available for decorative purposes. In use we also find pygmy and candle lamps. The life of these bulbs is quoted as 1000 hours and although they can be burnt in any position the optimum position is with the cap up. This is because the glass bulb may overheat in other operating positions.

The standard lamp with a coiled coil filament may have a bayonet cap (BC) or Edison screw (ES) fitting. They are generally available with power outputs of 40 W, 60 W, 100 W or 150 W and are found in bulkhead fittings or decorative and carriage luminaires. Bulbs rated at 25 W, 75 W and 200 W or at higher ratings are non-standard. The enclosures for the lamps are available in a wide range of fittings and sizes and can have a good resistance to both impact and adverse weather conditions.

Rough service lamps can be used if vibration or shock can be periodically encountered. They are constructed with a long filament so that more supports can be used to hold the filament but they have a slightly reduced lumen output. Tungsten GLS bulbs are typically filled with a mixture of argon/nitrogen to suppress evaporation and to give a higher efficiency. Certain versions, particularly the smaller types, use krypton instead of argon as this enables higher filament temperatures to be achieved. This leads to a higher lighting output.

The glass envelope may be clear or coloured, normally of a pearl finish. In clear lamps the actual filament is visible but pearl finish bulbs are preferred when the lamp is visible in use. The pearl lamp also features an internal acid spray on the inside of the glass bulb that helps the light to refract in different directions to create diffusion and effectively conceal the filament.

The advantages of the tungsten incandescent lamp are reliability, cost effectiveness and low power consumption but they suffer somewhat in that they have a restricted lighting level.

They also have an instant response to being switched on with immediate restrike capabilities and are connected direct to the mains with no control/starter gear being required. These bulbs are not recommended for use in external extended period lighting applications but are often specified for demand lighting since they have no inrush current or warm-up period so they provide immediate illumination. They do not contribute to any particular glare problem, are dimmable and do not create any measurable degree of light pollution particularly at the lower wattages.

Tungsten halogen lamps

These have been associated with security lighting for many years because they produce a very white instantaneous light together with good colour rendering. However, despite their wide use and the excellent service that they have given to the industry these lamps when employed within floodlights have also been installed on too many occasions in inappropriate circumstances. Although for vulnerable areas their good illumination properties can be used to great effect they must be mounted at a height sufficient to avoid glare and light pollution in other local areas. Equally the tungsten halogen lamp emits excessive heat directly in front of the luminaire and this can of itself produce a danger to persons close to the luminaire. These lamps also operate with an internal pressure above atmospheric and it is possible for the lamp to shatter so fragments must not be able to impact individuals. Unfortunately many budget price combined tungsten halogen floodlights with integrated PIR detectors have been commissioned in the domestic market but these have been mounted at a height to satisfy the sensor without considering the height that the floodlight is subsequently installed. In order to maximize the capture properties of the detector this has invariably led to the tungsten halogen lamp being installed at a low point and often being projected horizontally. A tungsten halogen floodlight should not be mounted at a height lower than 5 m, so if it is not possible to use this position in the installation and still obtain a satisfactory detection by the sensor, it is advisable to use a separate sensor and lamp. In the particular construction of the lamp itself a trace of halogen is added to the argon. At the point of the bulb wall this combines with tungsten evaporated from the filament and forms tungsten iodide which is transported to the filament by convection currents. A high temperature is formed near the filament with the tungsten being separated from the iodide and redeposited on the coil. The filament is itself operated at a high temperature and efficiency to produce an extremely white light source. The outer envelope of the tungsten halogen lamp is made from quartz so that the filament can operate safely at a far higher temperature than that of the GLS lamp and the pressure of the gas within the envelope can be further increased.

The linear lamps should be operated horizontally with a tolerance of + / −15° because at steeper angles the halogen vapour is forced to migrate to the low tube end resulting in a blackening at the higher end. It is necessary to prevent contamination on the outside of the bulb so they must be handled in such a way that grease cannot be deposited. This is because of the high temperature at which the bulb operates and if grease is present it develops fine cracks which obscures the light effect and causes lamp failure of the quartz envelope. Tungsten halogen lamps are supplied with a protective paper that is to be used when holding the tube and if touched by bare hands the bulb should be cleaned with methylated spirit.

Tungsten halogen bulbs for security purposes are normally of the traditional linear type being a double ended fitting and found in floodlights. Ratings are 100 W, 150 W, 200 W, 300 W, 500 W, 750 W, 1000 W and 1500 W. They produce a brilliant white light that is instant without a need for ballast or ignitor circuits. The floodlights may be enclosed or unenclosed with or without protective grids but all types must be mounted at sufficient a height to avoid glare and light pollution to neighbouring properties. These floodlights tend to have cooling fins to dissipate the heat and further increase the lamp life. Tungsten halogen bulbs give excellent colour rendition but unless quality units are purchased they suffer from extended bulb downtime. Similar to the GLS lamp it has instant restrike capabilities and needs no electrical/electronic starter gear.

In addition to the established linear bulbs there is a wide range of low voltage (12 V) lamps and PAR bulbs. The PAR bulbs are an option to incandescent reflector lamps and come in a range of diameters with ratings of 50 W and 75 W for security applications. These bulbs offer increased light output and reduced power consumption in relation to reflector lamps. They can offer either a spot or flood coverage and tend to be manufactured with ES caps. They are found in projectors that may have a number of holders to enable several lamps to be used within the luminaire and directed in different directions.

Tungsten halogen lamps are also dimmable but this can cause problems with the maintaining of the lamp wall temperature.

Reflector lamps

GLS and tungsten halogen bulbs are accepted in the mainstream security lighting market and are well known. However, reflector lamps also have a role to play but are rather different in that they are used in spotlighting an area or highlighting from a distance. They have a parabolic reflector and are available in a range of envelope diameters from 39 mm up to 125 mm. A variety of finishes is offered to include clear, diffused or translucent. Ratings are from 25 W through to 150 W and there are options on cap fittings.

These lamps are incandescent with a coiled coil filament and gas filled. The internal reflector that forms the beam is normally manufactured from pure aluminium and when evaporated onto the glass surface it produces a high level of reflectance.

If limited access is available extended life units can be specified. Extended life lamps save money by reducing the time that must be spent on maintenance. In addition fewer replacements mean less waste and disposal so this helps environmental issues.

Reflector lamps have a particular use to play in spotlighting internal areas in small commercial properties that are monitored by CCTV cameras. They give satisfactory colour rendering.

Certain PAR lamps also fall into the same class as reflector lamps but these have pressed glass reflectors. For security lighting applications they can produce a clear spot or clear floodlight pattern. They are good at spotlighting and highlighting at longer distances and are suitable for outside use. In these the envelope is made of heat resistant borosilicate glass with the coiled coil filament accurately aligned before the faceplate is sealed in position. A high reflectivity aluminium mirror coating is applied internally to avoid atmospheric deterioration. The ratings tend to be 80 W or 120 W using ES caps for fitting.

Fluorescent lamps

These tubes supply a high percentage of the artificial light generated in the UK. They are flexible in use because it is easy to alter the colour rendering, or the ability of a given light source to reveal the exact colours of an object and the characteristics of any installation, by changing the tube to a different variety.

A huge range of these tubes is available to cover many duties. They are popular since they are capable of generating the equivalent level of brightness as incandescent lamps but by using up to 85% less power. If consideration is given to the choice of lamp type and the control gear great savings in energy can be made. In practice they will be found to be in use in fittings extending from surface mount bulkhead size enclosures up to tubes that are a number of metres in length. They will also be found held in concealed sunk positions in ceiling voids so that they give off high levels of light yet are difficult to access by an intruder.

The assembled lamp consists of a narrow glass tube with the inner surfaces coated with fluorescent phosphor and principally filled with argon gas at a low pressure and containing a small amount of mercury. The tube ends are sealed with tungsten cathodes coated with a thermionic emitter. Caps are then fitted to the tube ends and it is coated in silicone to improve its starting capability in cold and damp environments. An inert gas is included in the lamp to further assist starting as the vapour pressure of the mercury is low.

There are a number of different phosphor technologies used from the traditional halophosphate to new generation triphosphors but all offer excellent colour rendering, high output and long life. These linear fluorescent fittings extend through a comprehensive range of formats to include different wattages, tube length/diameter, colour appearance, colour rendering, lumens and rated life.

The control of the lamp's current is achieved by the introduction of an impedance which may be inductive or capacitive. It can also be a combination of both so in some circuits capacitors are needed across the mains supply to correct the power factor.

There are a number of forms of control gear and switch starts used to energize fluorescent tubes including electronic circuits which are efficient, give instant start and are flicker free. The rated life can be further improved when operated on high frequency warm start control gear. All linear fluorescent tubes are intended for use over extended periods of time and although they may be switched on for shorter periods this is not ideal because of their start characteristics. They are not therefore used with short-term demand lighting but are ideal in security installations when illumination is required over long periods of time. They should be mounted in positions where they are not capable of being easily damaged by vandalism or abuse.

From this we may say that fluorescent tube lighting offers a solution in domestic, commercial and industrial applications. There are a range of interchangeable attachments to include diffusers, reflectors and louvres.

In addition to the traditional range of standard products there are weatherproof fittings available that also offer a high degree of resistance to the ingress of dusts and can be fitted with vandal resistant polycarbonate diffusers. These may be supplemented with vandal resistant screw fixings. There are also miniature versions that are still classified as linear but are actually circular in form and use a push-in cap installation technique.

The range of fluorescent fittings is enormous and has led to the expansion of the range of compact fluorescent fittings as follows.

Compact fluorescent lamps

These are valuable for commercial security lighting of low risk or for domestic applications. They have a small light source which matches filament lighting in both colour and quality but is less energy consuming. For this reason they are intended to be switched on for long periods of time as they present a reassuring lighting type that is cost effective. They are becoming more prominent in use because of their ultra compact size in the search for a means of miniaturizing the existing ranges of current luminaires. They use a triphosphor coating to enable the exhibiting of excellent colour rendering properties. Despite their ultra compact size they offer high output, high efficiency and excellent colour rendering by their cool white or tungsten white light form.

A huge variety of styles and types exist and they can actually interchange with coiled coil filament lighting by using adaptors to provide a quick conversion. They are expected to become even more widely used and to take away many of the roles presently carried out by demand lighting that uses tungsten halogen floodlights.

The range of compact fluorescent luminaires includes vandal resistant enclosures that are protected against the extremes of the weather. A respectable degree of protection can be afforded by mounting the luminaire at a height that is not accessible under normal circumstances to an intruder or vandal.

In view of the progression of compact fluorescent lighting (CFL) it is appropriate to identify the three most widely used forms of lamp:

- Single ended lamps with integrated electronic ballasts are offered as a direct replacement for GLS lamps. The integral control gear provides flicker-free start and substantial energy savings. These specify an average life of 15 000 hours for standard products and energy savings of up to 80%. They may be called retrofit as they have BC or ES caps. The envelope is available in a number of different forms and can even combine the classic shape of a standard light bulb.

- Two- or four-pin cap lamps are of a butterfly tube configuration. They provide an even light distribution over a large area and are often called 2D.
- PL lamps have a two-pin push-in cap. These have a distinctive double-turn tube and are used in luminaires with conventional control gear. They are slim and compact.

In addition to the three most used forms there are also four-pin bases for use on HF electronic control gear and for dimming systems. They are often used to replace conventional fluorescent tubes in modular luminaires. Triple-turn tube designs are shorter than double-turn versions so are ideal for compact luminaires. Flat format lamps with an innovative light source make small luminaire designs possible and are used in miniature luminaires when aesthetics are important.

CFL lamps therefore have a role to play in any security lighting application because retrofit lamps can be installed in existing GLS luminaires and then driven automatically by timers or photocells.

For new installations compact fluorescent lights are available in a wide range of forms to suit the different levels of security and to satisfy aesthetics when combined with amenity lighting.

We have come to appreciate that incandescent lamps such as GLS, tungsten halogen and reflector lamps with coil filaments rely for their operation on the passage of an electrical current through a metallic filament which when heated emits radiation. Discharge lamps, however, are different in their duty of operation as they rely on the passage of an electrical current through a gas or a vapour to function and produce light.

13.2 Discharge lamps

These fall into two categories namely low and high pressure but are all intended for use in areas that require lighting to be available for long periods of time as they all have ignitor circuits. Ultimately they are controlled by manual switches, separate or integral photocells or by timers. They are cost effective when used over long periods of time although certain judgements must be made in order to select the most appropriate type when colour discrimination is a need particularly if CCTV monitoring is to be invoked.

Discharge lamps do of necessity require electrical control gear to serve two essential functions. First, they must develop and supply a relatively high voltage to assist in the start of the lamp and then once the arc has been struck and established they must provide a current limiting function.

All discharge lamps do not satisfy their maximum light output at switch-on but need to go through a warming stage from their cold start.

The discharge lamps in which we are interested for security purposes are high pressure sodium, low pressure sodium, metal halide and mercury. These are all used to great effect for security purposes as they are extremely cost effective but can be run over long periods of time.

Traditionally low pressure sodium has been used if pure efficiency has been the requirement since it has an efficacy in the order of 200 lm/W but its output is monochromatic making objects appear as various shades of grey. If colour rendering is vital metal halide is a wise choice having an efficacy around 80 lm/W. As an option high pressure sodium and mercury give good colour rendering with stable colour over life.

High pressure sodium (SON)

This is the dominant light source for lighting over long periods and provides an extremely efficient golden white light source that allows colours that are being viewed to be distinguished. It is the most used light form in CCTV applications with extensive employment in security and amenity lighting areas with output of up to 130 lumens per watt. Typical ratings are 70 W, 100 W, 150 W, 250 W and 400 W.

SON lamps are widely accepted in industry for floodlighting and are manufactured to BS EN 60662. These bulbs have a long life and can be found in a number of variants to give even higher degrees of colour rendering. The essential lamp forms are:

- SON-E – elliptical. A coated elliptical bulb to minimize glare but giving marginally less light than the tubular version. These are dispersive and most suitable for industrial interior lighting where light quality is of importance.
- SON-T – tubular. A clear tube for floodlighting luminaires and supercritical photometric fixtures. These tubes are single or double ended and used in compact floodlights where size is important. Twin arc tubular lamps are used in areas where maintenance is difficult or expensive as they guarantee immediate restrike after a power cut.

SON lamps are not dimmable and if the electrical supply is momentarily interrupted a time delay is needed for restrike and there is a time delay for full luminous output to be restored.

The enclosures that form the luminaire housing come in many forms so can be matched to the installaton for aesthetic purposes or to give high levels of resistance to abuse. In addition SON floodlights can be installed at high levels so are difficult to access without the use of ladders or special equipment. Typical enclosures that are in use as an alternative to floodlight type housings are column globes, boulevards, top hats and carriage fittings at the top of lighting columns. Post lights will be found for illuminating walkways and wall packette housings are used in positions that tend to have used standard bulkhead fittings in original applications.

Low pressure sodium (SOX)

This is a high efficiency monochromatic light with low running costs. Its yellow light form is matched to the maximum sensitivity of the eye to give optimum vision at low light levels or in poor visibility but is unsuitable for applications where colour discrimination is required. As the light is monochromatic of orange/yellow form it means that they have poor colour rendering and most objects appear as various shades of grey.

These bulbs are manufactured to BS EN 60192 and have the highest efficiency of all discharge lamps with a classification of up to 183 lumens per watt. They are used predominantly in street lighting because of their high efficiency.

Low pressure sodium lamps are useful in conditions of fog or steam laden environments as the droplets of water in the atmosphere act as prisms and as the light is monochromatic the incident light does not become dispersed. Ratings are typical from 18 W to 90 W although some units are rated at up to 135 W.

The low pressure sodium tube can be identified by its relatively long length which is formed into the shape of the letter 'U'. They are installed horizontally and are to be held so that the limbs of the tube are arranged one above the other, as this produces the greatest light output.

These lights are not dimmable, are intended to be mounted at high levels and tend to have a more reduced line of options with regard to enclosures than those of the high pressure sodium alternative.

Metal halide

A light form originally used for the floodlighting of retail areas and industry. It produces a cool clear white light with very good colour rendering. These luminaires are used in those applications in which there is a need for both good colour discrimination and efficiency. Many of the enclosures of metal halide luminaires are similar to those of the SON light so there is a good range of options available.

The lighting tubes are of single glass envelopes and may be single or double ended with typical ratings of 70 W, 100 W, 150 W, 250 W and

400 W. They have an efficacy of up to 108 lm/W and are associated with the term MBI lamps.

Elliptical bulbs have a soft light distribution with less glare whereas the tube form is mainly found in low bay lighting.

Mercury lamps

These are a high pressure lamp and an alternative to conventional fluorescent lighting. They are generally termed as MBF lamps and provide a pleasing, cool, white light with adequate colour rendering.

Mercury lighting has for a long time been a popular choice for industrial interior lighting, utility projects and amenity lighting.

These were originally known as colour corrected mercury vapour lamps. A special type of mercury vapour lamp may also be encountered and this is referred to as a mercury blended lamp which includes a tungsten filament in its asssembly. This filament acts as a current limiting device and adds warm colours from the lighting spectrum from the mercury discharge.

Typical ratings are 50 W, 80 W, 125 W, 250 W, 400 W and 700 W with 1000 W tubes more specialist.

The bulbs are elliptical and may have an internal reflector. They feature a number of different cap type fittings.

Mercury lamps are not dimmable and their restrike capabilities involve a time delay because a time window is required to enable the lamp to attain its steady state luminous output. There are no restrictions on the operating positions of the lamps.

Mercury lamps do have a good range of enclosures available so can be used in wide ranging duties.

All discharge lighting fittings incorporate ignitor/starter circuits. These ignitors may be integrated within the lamp or be separate and held within the enclosure of the luminaire. It is important to install the correct replacement lamp in luminaires depending on whether an internal or external ignitor has been employed in the original installation. Symbols are used to provide guidance and these are shown in Figure 13.1.

A further guide of note is that related to common lamp caps since there are standard designations and IEC designations. These are as Table 13.1.

Classification of luminaires

The luminaire itself must also be classified against the relevant British Standard. BS 4533 is the standard that relates to luminaires (reference

Protection of persons against contact with live or moving parts inside the enclosure and protection of equipment against ingress of solid foreign bodies. Protection against contact with moving parts inside the enclosure is limited to contact with moving parts inside the enclosure that might cause danger to persons.	
First characteristic numeral	**Degree of protection**
0	No protection of persons against contact with live or moving parts inside the enclosure. No protection of equipment against ingress of solid foreign bodies.
1	Protection against accidental or inadvertent contact with live or moving parts inside the enclosure by a large surface of the human body, e.g. a hand, but not protection against deliberate access to such parts. Protection against ingress of large solid foreign bodies.
2	Protection against contact with live or moving parts inside the enclosure by fingers. Protection against ingress of medium size foreign bodies.
3	Protection against contact with live or moving parts inside enclosure by tools, wires or such objects of thickness greater than 2.5 mm. Protection against ingress of small solid foreign bodies.
4	Protection against contact with live or moving parts inside enclosure by tools, wires or such objects of thickness greater than 1 mm. Protection against ingress of small solid foreign bodies.
5	Complete protection against contact with live or moving parts inside the enclosure. Protection against harmful deposits or dust. The ingress of dust is not totally prevented, but dust cannot enter in an amount sufficient to interfere with satisfactory operation of the equipment enclosed. (Dustproof)
6	Complete protection against contact with live or moving parts inside the enclosure. Protection against ingress of dust. (Dust tight)

PROTECTION OF EQUIPMENT AGAINST INGRESS OF LIQUID	
Second characteristic numeral	**Degree of protection**
0	No protection.
1	Protection against drops of condensed water. Drops of condensed water falling on the enclosure shall have no harmful effect. (Drip proof)
2	Protection against drops of liquid. Drops of falling liquid shall have no harmful effect when the enclosure is tilted at any angle up to 15° from the vertical.
3	Protection against rain. Water falling in rain at an angle equal to or smaller than 60° with respect to the vertical shall have no harmful effect. (Rain proof)
4	Protection against splashing. Liquid splashed from any direction shall have no harmful effect. (Splash proof)
5	Protection against water jets. Water projected by a nozzle from any direction under stated conditions shall have no harmful effect. (Jet proof)
6	Protection against conditions on ships decks (deck watertight equipment). Water from heavy seas shall not enter the enclosure under prescribed conditions.
7	Protection against immersion in water. It shall not be possible for water to enter the enclosure under stated conditions of pressure and time. (Immersible)
8	Protection against indefinite immersion in water under specified pressure. It shall not be possible for water to enter the enclosure. (Submersible)

Suitable for use with external ignitor circuits.

Lamp incorporates an internal starter.

Figure 13.1 *Symbol guide for discharge lighting and degree of protection (IP)*

Table 13.1 *Common lamp caps*

Standard designation	*IEC designation*	*Description*
BC	B22d	Bayonet cap
SBC	B15d	Small bayonet cap
3-pin BC	B22d-3	3-pin bayonet cap
ES	E27	Edison screw
SES	E14	Small edison screw
GES	E40	Goliath edison screw
–	GU5.3	Bi-pin
–	R7s	Recessed single contact
–	G4/GU4/GY4	Small bi-pin

can also be made to EN 60598). In this details are given as to the protection provided against electric shock, ingress protection and the material of the supporting surface.

Electric shock

Equipment is to be given a protection type safety class and this is to be shown on the type plate. This classification will apply when it is properly installed (see Figure 13.2).

These exist as Class 1, 11 or 111 depending on the level of electrical insulation that the luminaire affords. This is initially claimed by the manufacturer of the goods in accordance with the construction of the product. These safety classes are:

Safety class I.

Luminaires are provided with an earth connection.
The earth conductor terminal is marked.
Under fault conditions the power supply is cut off by an overload device.

Safety class II.

Completely insulated luminaires. Together with the functional insulation these have an additional protective insulation. Under fault conditions no dangerous voltage can reach metal parts which can be touched. Many luminaires are available in Safety class I and also in safety class II.

Safety class III.

This covers luminaires operating at extra low voltage. They may only be connected to safety insulating transformers according to EN

60742/VDE 0551 or another power source according to VDE 0100. If an insulation fault occurs no problem high–touch voltage can develop.

Ingress protection (IP)

This is the classification of the level of protection afforded against the ingress of solid objects, dusts and liquids. The designations for the selection of the ingress protection are listed in Figure 13.1.

Material of the supporting surface

BS 4533 includes information regarding the materials for the supporting surface of the luminaire, taking into account flammable and non-flammable surfaces.

This information is as the extract shown in Figure 10.1 from the emergency lighting systems section.

The luminaires should also show compliance with the EMC Directive (89/336/EEC) and the Low Voltage Directive (73/23/EEC) and carry the 'CE' Mark.

In addition to the normal ways of upholding the classifications of luminaires in an installation sense it is possible that manufacturers' data sheets may carry additional instructions. These instructions are to ensure that the installation is carried out correctly and safely and that the luminaire is used and maintained effectively. Such information is to be given on the luminaire, on a built-in ballast or on separate sheets provided with the fitting.

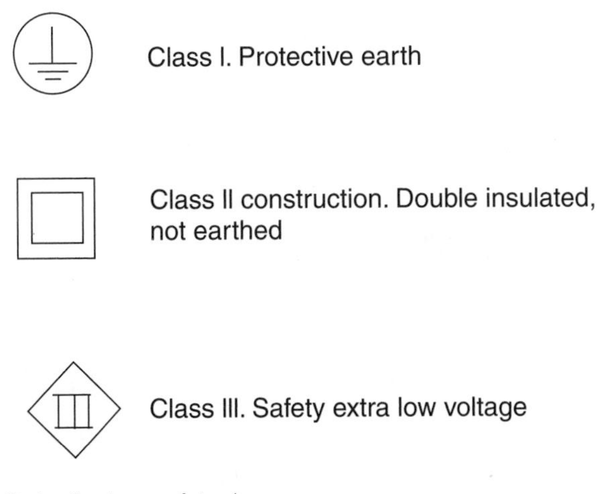

Figure 13.2 *Protection type safety classes*

Good examples are tungsten halogen floodlights and discharge lighting luminaires as these run at elevated temperatures and need certain points on their enclosures to be isolated by air gaps from such items as pipes and fascia boards etc. that will be attached to the main fabric of the building.

We are now at the stage at which it is possible to identify the key issues in order to determine the luminaires to be selected for the specified duty. These considerations are to be largely based on:

- Light level required.
- Need for colour discrimination.
- Exposure to the weather.
- Level of risk and probability of vandalism.
- Optimum position for mounting of luminaire.
- Demand or long-term lighting required.
- Support by other security technologies or manned guards.
- Lighting to be integrated with other technologies and building management.
- Light pollution and light overspill to adjacent or local residential areas.

The market for security lighting is healthy so there will always be a good range of luminaires available for the engineer to choose from.

The innovative engineer may also want to address further the problem of light pollution by selecting floodlights specifically designed to suppress the nuisance effect. These fittings recognize the need to avoid upward lighting scattering dust particles and obscuring the night sky. In addition they resolve the problem of light trespass in which light falls outside of the area that it is intended to illuminate causing a nuisance to neighbouring properties and glare to others.

Flat glass panels that help to stop light above the horizontal when the lamp is directed downwards have always helped. Unfortunately if floodlights are aimed incorrectly and they have badly designed optics the light will continue to overspill.

Developments with new generation floodlights featuring special optic systems and designed for use with the glass horizontal are now capable of defining the light path and these cut off sharply the areas that would otherwise cause light trespass and spill. They also address the issue of sky glow and optional lights can actually be used with confidence for facade lighting. These project from a ground level and ensure that the lighting terminates at the top edge of buildings.

In other respects surveyors are governed by the quality of the goods that they select. Luminaires will be constructed typically of metals, glass and plastics other than in specialist applications. These methods of construction will essentially be time honoured and produce products that have been amended, as a learning curve has been generated in the market place. The range of luminaires available to us is without doubt very much diverse and offers a genuine and realistic solution to every security lighting application.

14 Detection techniques

Although we are all familiar with manual switches that are used to switch on lighting and to control system networks the normal technique in the security lighting industry is to use automatic technologies. This ensures that systems are brought into effect automatically so that it is not necessary to rely on the actions of an individual to energize the lighting. Manual switches tend only to be used as a support item to override automatic practices.

Extended period lighting tends to be activated by timers and photocells but demand lighting is energized in response to the activity from an automatic detector or sensor. In practice an automatic sensor carries out the role that a switch would perform manually. Sensors are also used in energy management systems as occupancy detectors to bring on lights when areas are occupied and then to monitor that area until it is vacant. At that point the lighting would be turned off or reduced in its level of illumination.

14.1 Automatic detection

Even though there are specific detectors in common use with security lighting, in practice any recognized sensor can be employed to automatically energize the luminaires. This detection is that of sensing a person or vehicle. In all cases the detectors have a maximum switching capacity to control the lighting directly but the detectors can be used in conjunction with contactors or relays so that the load is taken through these devices' main switching contacts.

Sensors can be mains powered or be extra low voltage and come complete with a particular controller that will provide options on the operation of the lighting system. These extra low voltage sensors are useful in applications where it is difficult to install heavier mains cables and satisfy the more onerous protection techniques demanded of them by regulations.

In some instances the detectors will be very much overt so that they can be easily seen. However, in other instances they are hidden and intended not just to bring in security lighting but also to give outputs to other systems and signalling devices in order to warn of the approach of persons or vehicles.

The most used technologies for automatic detection are:

- Passive infrared
- Active infrared
- Microwave
- Fluid pressure
- Electromagnetic cable
- Fibre optic cable
- Capacitive field
- Geophones
- Piezoelectric

Passive infrared (PIR)

These are the most adopted external detector types used in automatic security lighting systems particularly in low risk situations that also include amenity lighting. They are often found as the switching element in occupancy detectors used in building management sensors to recognize the arrival of people and their continuous presence.

These detectors are passive in the sense that they do not emit energy but are receivers of far infrared rays. The optical window of the sensor catches IR and focuses it onto pyroelectric sensor elements. These elements then absorb the IR and transfer it into heat. When the amount of IR energy they receive changes, the elements themselves change temperature and this leads to the creation of an electrical signal. The heating and cooling of the pyroelectric sensor generates the signals that are analysed by the detector.

A single pyroelectric sensor will generate signals in response to any change in IR levels so they are sensitive to background temperature changes, draughts and other environmental disturbances that can be a problem and lead to sporadic operation in some environments. This is overcome by adopting dual opposed sensor detectors which combine two elements each with opposing charges within one detector. Each of these elements creates a signal (either positive or negative) when it changes temperature but when both are affected simultaneously the positive and negative charges cancel each other out. In normal use an infrared source must appear in an active zone on one section and then appear on the other within a given time scale or time window. Using a time scale pulse count the alarm relay will only activate to a number of trigger strength signals. An extension of this technique is to use quad sensors which effectively have two dual element outputs fed to a signal processing unit which will only go into alarm when two signals from the quad system exceed a predetermined threshold. From this we can say that the PIR detector activation is the result of the radiation of IR energy reaching a pyroelectric element.

The PIR sensor is used extensively as a standalone device that can detect the infrared energy emitted by a human target or from the heating effect from a vehicle and can then trigger the lighting directly or through a controller. These PIR sensors are often also found as the sensing device integrated with standalone luminaires. However, in general it is advisable to mount the luminaire and the sensor separately so that they are both sited at their optimum positions rather than seeking a position of compromise that partially satisfies the two essential items.

For low security risks there exists a huge range of floodlights that have an integrated passive infrared sensor but these must be mounted at a height to avoid glare and this has a major influence on the coverage pattern then afforded by the sensor.

The PIR detector remains widely used as a mid-risk security device in the protection of open areas. They are efficient, inexpensive, reliable and give high value aesthetics. The PIR, however, must not be able to view pulsating heat sources such as from extraction and heating units. It is also troubled by animals moving into its range of cover or from birds moving close to its sensor element. The passive infrared device is more sensitive to movement across it than away from it and should not be subject to direct light. In addition they are not to be mounted above a luminaire because the effect of the heat rising from the lamp would cause the sensor to give sporadic operation.

The PIR used for external detection systems can give a range up to 15 m volumetrically. They are available in a multitude of patterns to include long range up to 30 m and can equally be used in a parallel circuit with beam interruption devices to help filter out false alarms. This is achieved by wiring the devices so that both must alarm simultaneously.

The PIR is available in many diverse forms because of the wide use of this sensor technology. It has long been used as an internal detection device in intruder alarms and is capable of stable and consistent operation. However, when used as an internal detector it is mounted in areas that have environments and areas that can be more clearly defined. When passive infrareds are used externally they are subject to extremes of weather and the areas in which they are operating are more readily affected by problematic sources such as sunlight, car headlights and wind so it is more difficult to filter out all false alarm hazards. Nevertheless we do have a great range of devices available to us for both indoor and external use and to adjust the units for length of coverage and range.

We will therefore see options available to us to include wall or ceiling mount versions with volumetric, long range, corridor, curtain and up to full 360° patterns. Full adjustments can be made available by pan and tilt heads with high impact housings giving the necessary level of protection to difficult environmental conditions. There are customized optics available with lens libraries in order that the cover can be easily amended or blanking pieces can be added to blind parts of the sensor from looking at false alarm sources.

The electronic circuitry can provide sunshine compensation plus RFI and EMI immunity with digital temperature compensation, automatic background levelling and UV stabilizers.

More complete sensors will be found to have CCTV or video outputs integrated in the units with 24 hour contacts that may be for normally open or normally closed switching.

The object is to select the most efficient detector to satisfy the application. False alarms can be reduced by using pulse counts. These techniques use a counter to record a specified number of trigger strengths within a given amount of time before activating the alarm relay. If the required number of signals is not generated within the desired time frame the counter returns to zero. These pulse counts are adopted now by many detectors as standard and can be adjusted on site to reach a compromise between high levels of capture and reduced false alarm conditions.

When used in internal applications PIRs can virtually be guaranteed not to generate false alarms but in external applications this cannot be guaranteed although most manufacturers now claim that due to advances in technology PIRs are less prone to false activation by the wind or rain etc. However, the products are not to be situated near to heating and ventilation ducts or exhaust fans because by their very nature they will cause false triggering.

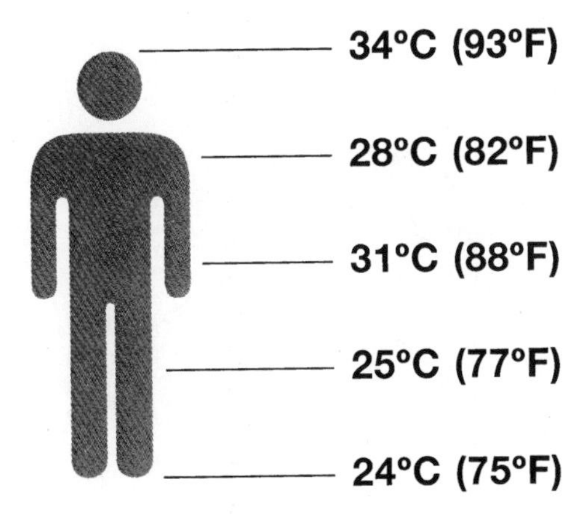

Figure 14.1 *Human body infrared radiation levels*

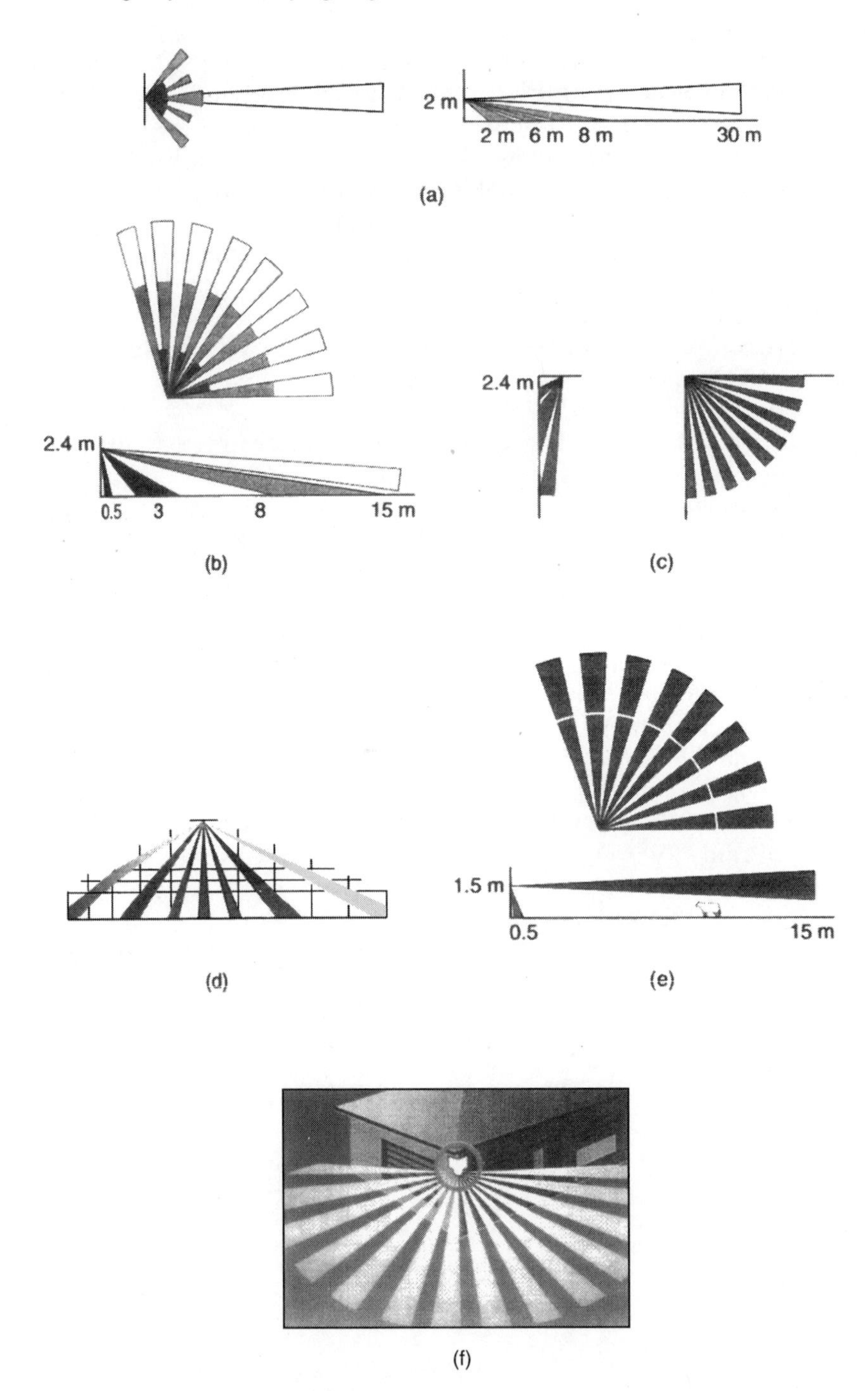

Figure 14.2 *PIR sensor coverage patterns (typical). See text for explanation*

Nevertheless if it is vital to check the validity of each and every signal this can be achieved by using verification and CCTV monitoring. However, these are other detectors available to us that may be more suitable or can be used alongside PIRs in a dual role to negate any false alarm activity.

Before concluding our investigations on passive infrared sensors it should be said that every object actually has a measurable temperature. The human body recognizes infrared radiation as warmth as this is part of the spectrum of electromagnetic waves. Figure 14.1 depicts the levels of the human body. Figure 14.2 illustrates the most popular sensor coverage patterns available using passive infrared techniques.

Figure 14.2(a) illustrates a long range pattern with additional cover close to the sensor head. The standard volumetric cover is shown at Figure 14.2(b). The more specialist curtain pattern appears at Figure 14.2(c). This pattern is used to retain the coverage close to a wall or perimeter. Figure 14.2(d) shows a ceiling mounted detector that provides an essential 360° level of protection. Figure 14.2(e) shows a blanked area to avoid animals or pets. A popular 180° protection pattern and mounting technique are illustrated in Figure 14.2(f).

Active infrared

This is known as a beam interruption detector and in operation it can cope very well with extremes of climate and temperature.

The detector comprises two essential components, namely an infrared IR light transmitter and a receiver. When an object interrupts the signal between the transmitter and the receiver the alarm output is energized. It operates in the region of 900 nm wavelengths at a carrier frequency of 500 Hz using an IR LED within a protective enclosure that is filtered to operate at the correct frequency. It is pulsed rapidly to provide a concentrated beam of IR that does not generate heat. The receiver is contained within a single chip using a photoelectric cell to transduce the energy to hold the alarm circuit in a quiescent state. It is therefore an active device as it actually transmits IR energy unlike the passive sensor that only collects it from objects in its field of view. The infrared energy is modulated to prevent the receiver being affected by another source and to stop the receiver responding to an incorrect IR signal.

Synchronization techniques are used to ensure that the transmitter can only operate with its correct receiver. The system is set so that the beam is monitored using time-based multiplexing.

During the installation process the photobeam alignment is achieved by status communication with the beam alignment level usually being visually displayed using LED indicators on both transmitter and receiver. The alignment status at the receiver is optically transferred to the transmitter with the lenses adjusted for alignment by designated adjustment screws for the vertical and horizontal levels. The operating range is generally quoted to be in the order of 200 m but in stable internal applications the distance can be as great as 600 m.

If there is a possibility that an intruder could step over the beams tower enclosures can be used to stack the beams to a greater height. In the event that the beams are to be installed in an area that is subject to fog, automatic gain control (AGC) is normally specified, as this adjusts the trigger level in response to environmental changes. Heaters are used in cold environments to clear the beam windows and power supplies are upgraded to handle the additional load. The modern IR sensor is essentially stable in operation as it employs multiple beam paths that must be broken simultaneously so debris or birds going through the beams do not create a problem.

To ensure good detection and to avoid false alarms the only considerations are:

- Avoid siting in areas that can suffer from strong sunlight or from headlight glare direct on the transmitter/receiver line.
- Do not install where the windows can be splashed by dirty water.
- Do not site the equipment on unsteady surfaces.
- Regularly check for obstructions between the transmitter and receiver.

The active infrared detection system has particular uses when the beam is projected across a gate, barrier or opening or is run inside of and parallel with fences or walls. It can equally be run on top of fences and walls to stop attempts to climb over them.

The enclosures are robust and weatherproof so can be used with confidence in areas that are open and exposed to difficult climatic conditions.

The controllers used to drive the system have relays and volt-free contacts to enable the detectors to interface with any other electrical system or a number of different signalling mediums.

Microwave

These sensors are used as a beam interruption device in the same way as the active infrared.

The microwave sensor radiates its beams of microwave energy from a transmitter to a receiver that is in direct line of sight. The receiver compares the energy level with that transmitted and if a significant difference in wave form, amplitude or magnitude is found for a given time window, it goes into alarm. This is in contrast to the Doppler effect microwave that is used in internal applications. The properties of microwave beams are governed by the type of transmitter that it uses although in general they have a greater depth of coverage than the IR beam.

The field of microwaves can drift so they are best employed within a boundary or physical fence to ensure that only unauthorized access is detected otherwise the coverage may drift to adjoining areas that are not intended to be monitored.

The microwave can provide a secure area up to 200 m and perform well on long, straight boundaries but in those situations where there is ground undulation then beam shaping or phase sensing is to be used. This is achieved with aerials to give increased ground and high level cover. The type of transmitter selected is related to the area that it is to protect. Vertical patterns have an energy form based on a parabolic trough which is extremely broad. When pole mounted they are sited at only 1 m above ground level. Short-range detectors can be used to cover gates with a cylindrical pattern. Long-range beams can extend to 300 m but beam width is diminished and an option is to install a number of mid-range sensors.

With respect to false alarms although the microwave is not affected by fog, rain or frost it is a highly sensitive device so it can be easily troubled by the movement of grass, bushes and trees. In addition it is not possible to determine its exact range in the same way as the IR device and its pattern can drift considerably. For these reasons it is unable to give an equivalent stable operation to that of the active infrared. Therefore it tends to have its activation proved and supported by CCTV systems.

Fluid pressure

This sensor type is a further device used to protect boundaries.

It comprises a small diameter tube sealed at one end and with a flexible wall that if compressed causes pressure to be applied to an enclosed fluid. This change is sensed for magnitude and the differential is converted to an electrical pulse. The device responds to a person or object applying pressure upon the ground in which the tube is sited. The installation may consist of one or a number of loops with multiple loop systems analysed for pressure changes over a large expanse.

The tubes are located in trenches under an elastic surface such as gravel, sand, asphalt or grass. It is difficult to embrace installations under concrete or solid surfaces.

False alarms are normally only a result of ground movement or from the foundations of objects in high winds.

Electromagnetic cable

This is a cable type sensor that can be used as an option to the fluid pressure sensor.

The detector cable has an inner and outer conductor that are themselves separated by a dielectric filler. When a current is applied to the separated conductors an electromagnetic capacitive field is created but this is interrupted by the dielectric layer. A change in the outer conductor is caused by the application of external pressure and this transforms a mechanical signal to an electrical pulse. The amplitude of the pulse is in proportion to the external force in that it responds to vibration, cutting or interference.

The installation of the cable is generally by fixing to the inside of a fence or perimeter being protected. The fixings are related to the actual cable type and the greater the number of fixings, the more efficient the system. These are also at times referred to as a microphone and found buried in gravel.

The incidence of false alarms is governed by the sensitivity setting of the processor and its ability to deal with environmental disturbances.

Fibre optic cable

This form of cable is mainly used in hazardous areas where all electrical systems are to be intrinsically safe. Fibre optic transmission is achieved by transmitting a beam of light along an optical fibre by internal reflection along the cable core.

The receiver of the transmitted light responds to a change in magnitude of the received signal as a result of interference or cutting. This then creates the desired signal.

The fibre optic cable is best contained in the hollow core of the perimeter that it protects as it is easily damaged. The cable comprises a core or filament of extruded glass in a continuous length. This glass is essentially pure with a high transmissivity to light along its length. The glass fibre is surrounded by a sheath that provides mechanical protection for the fibre and is usually referred to as a jacket.

Resistance to false alarms is governed by the quality of the equipment and the transmission/receiver compatibility although the cable needs to be cut to create a positive alarm. It is therefore used in high security applications and to support other technologies.

There is a growing recognition of fibre optics within the security industry for many applications and particularly in harsh environments and in long cable runs where the cable can be protected. The advantage of fibre optics is that all signal transmissions are possible and there is little loss (attenuation) through the cable itself. Amplification is not therefore needed and electrical or radiated energy interference is not possible. In addition there can be no corrosion and the cable can withstand harsh environments. It suffers in that it is an expensive medium in relation to the more traditional types and care must be exercised in bending and terminating the cable.

The increasing use of optical fibres will inevitably lead to a greater use of this cable type to provide a wider use of security functions.

Capacitive field

This technique employs electrical resonance in that the detector senses a change in current caused by an appropriate change in inductance or capacitance by a person entering the detector's capacitive field. It may also be caused by touching its resonant tuned circuit.

Capacitive field sensors are available in cable form. They may be applied on fence top posts to carry a low voltage providing a capacitive effect between the wire and a grounded post. As an alternative they can use the surrounding air to store a charge. The system processor analyses the magnitude of change to assess the probability of cause. False alarms tend to be the result of objects being blown into the protected area or by driving rain or snow.

Geophones

These are devices made up from coils with a magnet held in a certain orientation by spring tension and creating an electromagnetic force. If the magnet is displaced by positive movement or pressure a current is generated in the coil and analysed as an alarm.

The installation is performed by securing the device to a fence, wall or any other protective medium that would be subject to cutting or sawing. They are also at times found buried.

The surface to which the geophone is mounted should not normally vibrate otherwise false alarms will occur and if buried they are not to be in an area surrounded by trees.

Piezoelectric

These detectors respond to mechanical stress, strain or compression to produce a proportional voltage. The piezoelectric coefficient is its defined ability to determine to what degree the mechanical stress has been exerted.

Such devices are installed as cables so as to recognize mechanical strain via an analyser.

The piezoelectric device is surface mounted or buried to a clearly defined manufacturer's specification to control false alarm rejection.

14.2 Dual technology verification

In the electronic security industry there is an ongoing process of events leading to the reduction of false alarms particularly from intruder alarm systems. However, security lighting is very much a separate subject and because of the very nature of the use of the detectors there can never be a means of always ensuring that demand lighting is only energized in response to a definite and wanted activation.

On the assumption that a high degree of sporadic operation does not occur on a regular basis we may conclude that false activations that bring on lighting cannot be construed as a problem providing this does not cause light pollution or distress to neighbouring properties. Nevertheless there must be an exercise carried out with the aim of achieving demand lighting being energized only when called upon.

The most effective way of removing false signals that have been caused by environmental disturbances is to use two or more sensors that use different criteria for their operation. These are wired in parallel circuits so that both must respond simultaneously to create a signal at the lighting control equipment. As an example the response from a passive infrared sensor would be disregarded unless ground movement is also detected by a fluid pressure sensor at the same point in time.

The intruder alarm sector of the security systems industry has a number of dual technology detectors available to it. These sensors use a number of detector technologies in one enclosure but they are normally used in internal applications so could not have a major role to play in the mainstream security lighting area. This leads us to say that in real terms the nuisance value of unwanted spells of demand lighting is to be balanced against the cost and complexity of introducing greater levels of technology and additional costs.

Although we have covered the main sensor technologies that are used throughout the security lighting industry for the different levels of risk, it will be found that there are also a number of devices used in the domestic low security field to provide an amenity function, but that do offer a small degree of security. These are also at times amalgamated with other technologies.

As an alternative to using passive infrared techniques ultrasonic sensors using the Doppler effect or radar technology pattern are integrated in lamps used for internal applications. These sense movement so can respond to a door opening to give an automatic light. They have an adjustable lux setting together with a timer and are used to bring on a light at the point a door begins to open and prior to a person entering an area or room. This gives an element of hands-free operation because no manual action to put on a light needs to take place. It is often combined with access control systems that have proximity tags that are worn by a user and signal a door to open when it is approached by an authorized person.

Remote controlled IR switches with dusk to dawn photocells can be used to energize lights from a safe distance of up to 10 m and portable wireless pagers can detect signals from security lighting detectors. As an extension to this idea radio transmissons can be used to link detectors to plug-in wall outlet receivers and chimes that in turn switch on lights. These methods are all used to expand the range of options open to us for the different levels of risk and to overcome difficulties with cabling in difficult areas.

The technique of installing mainstream demand lighting follows in the next chapter.

15 Demand lighting

This form of lighting is only switched on for short periods of time under normal circumstances, although it is often provided with manual overrides so that the luminaires may be energized for slightly longer time spans as and when the need arises. Examples are lights that are selected by detection devices to illuminate and protect a vulnerable area that is liable to be attacked under cover of darkness but may be used as an amenity point at other times. In the domestic environment a vulnerable patio door can be protected by a PIR tungsten halogen floodlight mounted at a high level above it and this luminaire can be wired with a manual override so that the lamp can also be used for amenity purposes. In the industrial sector floodlights may only be wanted for security and over short periods of time so they can be controlled in a similar way. However, a facility should exist in order that they can be switched on for greater periods in the event that a guard wants to patrol an area for a slightly longer time scale than is normal. The most practised technology of bringing on lights automatically is by PIR sensor technology. This forms the nucleus of the security lighting industry.

We will recall that the PIR sensor, which is often called a motion detector, is the foundation for the domestic and low security risk application sector. It can be used in conjunction with almost any light type to match the surroundings of a home or commercial practice in which aesthetics are also important. It detects sudden temperature differences of heat emitting moving objects such as people and vehicles by measuring and comparing infrared radiation within its detection zone. The human body recognizes infrared radiation as warmth as this is part of the spectrum of electromagnetic waves. Indeed every object has a measurable temperature whereas the objects have a tendency to compensate temperature variations. Sudden temperature variances only are detected. The built-in twilight zone setting that is combined with the sensor ensures that the lights are only switched on when natural light is not available and when motion is sensed, so this is convenient, deters crime and saves energy.

There is a vast array of lenses that are used by PIR detectors and by turning the sensor sensitive areas can be pinpointed and the sensitivity can be adjusted. The range may also be shortened or lengthened by tilting the unit vertically and shrouds are at times made available so that areas of the normal pattern can be removed if they overlook an area that is not to be viewed.

The twilight setting is generally set by a control that is rotated between a brightness of 1–100%. This can enable a normal light intensity of, for example, 20 watts to be switched to 100 watts when approached by persons. The rating of the detector will be specified by the manufacturer but in the main they allow for some 1000 watts to be switched with additional loads requiring an intermediate relay or contactor.

The cabling to a mains powered sensor is straightforward being only governed by the IEE Wiring Regulations and standard codes of practice in the same sense as standard wiring.

The connections to the detector are as shown in Figure 15.1. This is easily understood in that once the sensor element of the detector is triggered the internal relay of the device switches the load.

In the event that a manual override is needed this is effectively wired in parallel with the relay contacts.

If a light already exists and it is intended to make this automatic in operation it is often possible to add a sensor to the installation at an acceptable position and then bring the switched live connection into the original luminaire. In other applications it is better to retrofit a combined luminaire with a passive infrared sensor in place of the original light fitting. It is to be accepted that in the latter examples it may not be possible also to have manual overrides unless additional cables are added to the installation.

In Figure 15.1 we take account of a mains powered sensor but it will be found that there are also light control systems that although they feature PIR sensors they use a mix of mains wiring and extra low voltage cabling. In these systems the sensors are extra low voltage, normally 12 V. Such systems are most used when installing certain parts of the wiring to some of the sensors or ancillary equipment would be difficult so extra low voltage components are used. These are wired via signal cable as found in intruder alarm or communication practices. These light control systems can consist of two main sections that are secured together. First there is the control or junction box which is used to mount and connect the unit. It will contain the daylight sensor, control timer circuitry power relay and the terminals used in the wiring. The other main section is the sensor head that can be moved on its adjustable junction and is used to house the PIR sensor that operates from the in-built extra low voltage power source. With these scanning and automatic switching device systems the power relay is brought into operation by an electronic timer circuit which determines the interval for which the lights stay on after detection of movement.

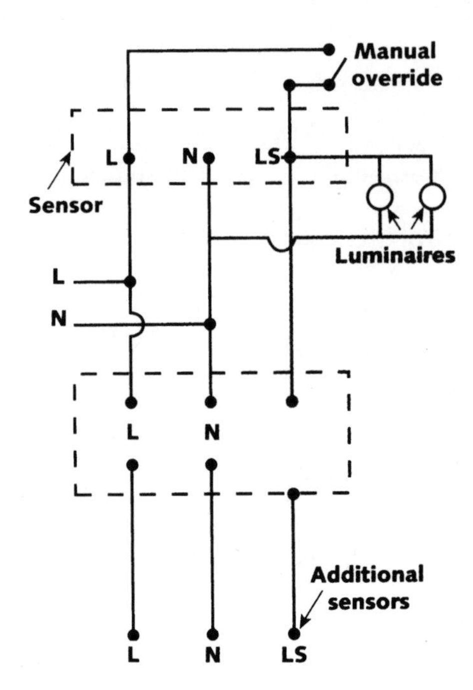

Figure 15.1 *Mains powered PIR sensor wiring*

In view that the system resets at every detection the connected lights will stay on for as long as the person remains within the field of view. This time interval is adjustable but a walk test facility is always provided to reduce the time interval to a few seconds. This allows expedient checking of the field of view during installation so that adjustments may be made.

Daylight sensors are also incorporated so that lights are only operated during the hours of darkness but an option is usually provided so that they may also be energized during daylight conditions if required. Naturally a manual override should be provided so that the lights can be operated from a remote switching point. The wiring of this must very much depend upon the system that is being employed but an example of automatic operation with manual override is shown in Figures 15.2 and 15.3 employing live, neutral and load terminals. Installation is straightforward; mains power is supplied from a fused source and taken via a conventional wall switch to the unit. The switch is an important part of the system as it provides control of the functional operation, the supply cable then being run to the luminaires. One should appreciate, however, that, the supply should remain in an active state and should not be switched to control the load.

With these automatic switching device systems the wiring is taken up to the control box but remote switching can also be employed. A variant on this is the type that comprises of a sensor and a separate controller that is mounted inside the premises. The controller, which comes complete with an integral transformer, is mounted indoors and connects to the main power supply in place of a light switch. The transformer circuitry supplies the extra low voltage to the sensor and the controller monitors the state of the detector. Use controls can be switch 'on', 'off' or 'automatic'. There is also an adjustable 'on time' switch to select the desired lights delay once the controller has been triggered. With the switch in the on position the lights will be illuminated at all material times and with it in the off position the lights are extinguished at all material times. However, with the switch in the automatic mode the lights will come on during the hours of darkness when the sensor or sensors have been triggered and remain illuminated for the period set by the adjustable 'on time' switch.

The controller display can show the state of the installed luminaires and the occasions when they are illuminated by an onboard screen.

The sensor heads are weatherproofed to an appropriate IP classification with full coverage adjustment. It is possible to connect a number of sensors to one controller and these may have different zones and select different lights or they may be programmed so that any sensor can select the full lighting system. This is governed by the type of controller and system used but in all cases a photocell in the sensor head automatically cancels any daytime operation but this is generally an option in the event that the network is to function in a 24 hour mode.

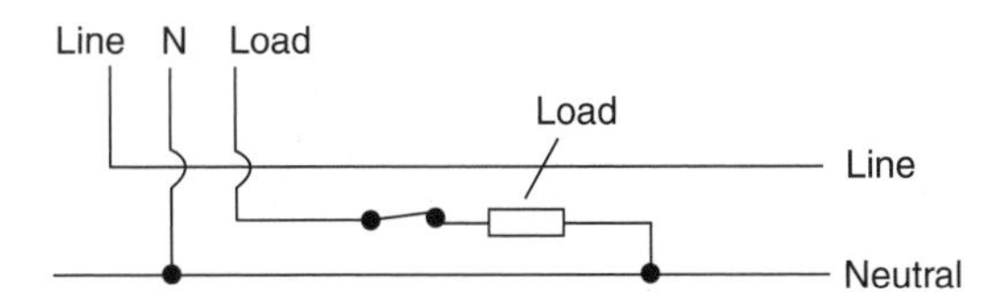

Figure 15.2 *Manual override off plus automatic*

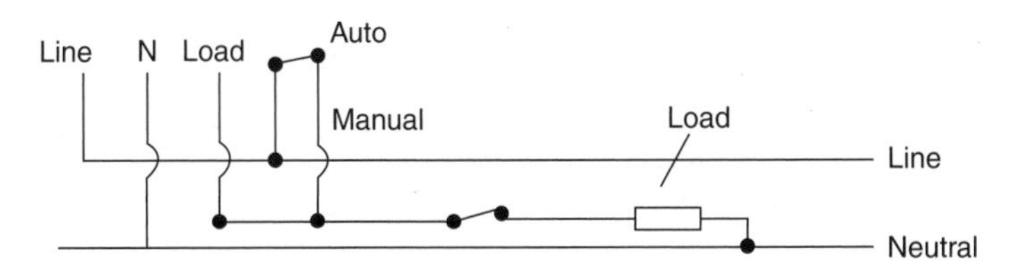

Figure 15.3 *Manual override on off plus automatic*

Other devices such as bell pushes or switches can be used to trigger the lights. These may be normally open or normally closed depending on the exact system but they allow a user to energize the lights without having to move the controller from its automatic setting. In practice the bell push presents a signal across the zero volts and the trigger terminal so it manually simulates the operation of the PIR detector. On leaving the premises the operator can operate the bell push in order to activate the lights hence they are immediate and a user need not have to make a number of steps into the outside of the premises and activate the sensor before lighting becomes available. The countdown cycle time will not commence until the protected area is finally vacated.

Figure 15.4 illustrates a light control system with extra low voltage sensors.

It becomes apparent to the reader that demand lighting can be started up from any form of automatic sensor although the passive infrared is by far the most used. For this reason the control equipment that is used in conjunction with this sensor type is readily available and external PIR sensors come with time-honoured and easily understood data to enable them to be wired to any existing or new lighting system. Nevertheless all of the sensors covered in Chapter 14 may be used to trigger lighting although it is possible that they may need additional power supplies or interfaces such as relays to control the loads.

In order to exploit other building management systems or security systems there are a number of ways that lighting can be integrated with these other technologies and the capacity and flexibility of any system can always be extended by the engineer being alert to variant techniques. Although there must be a limit to the size of the actual lighting system, there are a number of ways in which it can be easily integrated or customized.

Pneumatic time switches can be used to interface with mains voltage sensor heads enabling the luminaires to be pulsed before exiting a

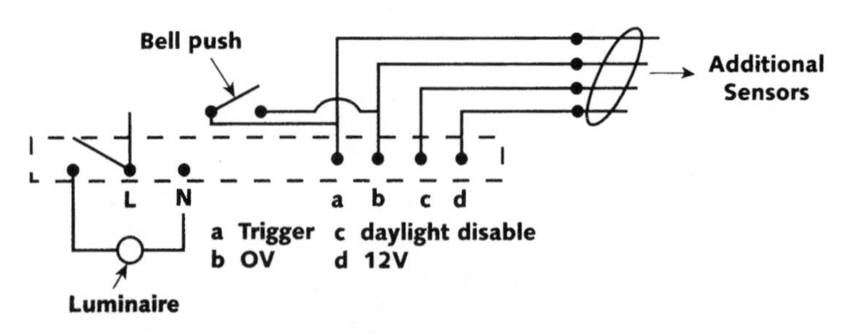

Figure 15.4 *Light control system with extra low voltage PIR sensors*

premises. This provides the same feature as the extra low voltage controller's technique that can signal the lighting via a momentary switch or bell push. It allows the user to trigger the luminaires simply by touching the switch so no steps need to be taken into a dark area before the lighting is made available. In effect, the push switch manually performs the same function as a sensor head does automatically, and can have the same lights-on time period (see Figure 15.5). The wiring is different to a traditional two-way switch as shown in Figure 15.6 because the sensor head and pneumatic time switch reset automatically.

If there is a need to have an indication of lighting being in operation, this can be achieved by placing a neon indicator across the actual lighting load or with a 12 V system by using an LED and wiring a 1 k resistor in line. These methods are depicted in Figures 15.7 and 15.8.

Relays can also be used to link security demand lighting to any alarm output with the time governed by the output switch period. A classic example is a bell shut-off time when used in conjunction

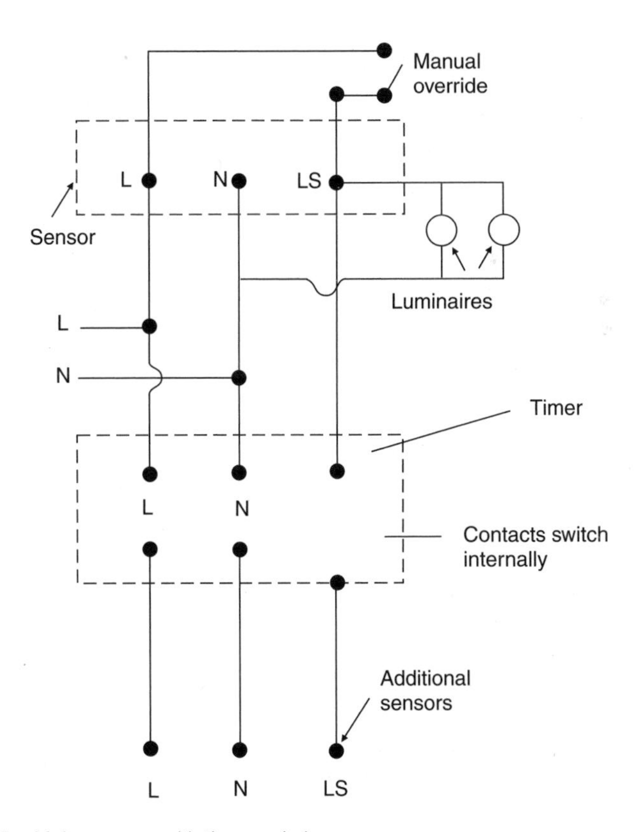

Figure 15.5 *Mains sensor with timer switch*

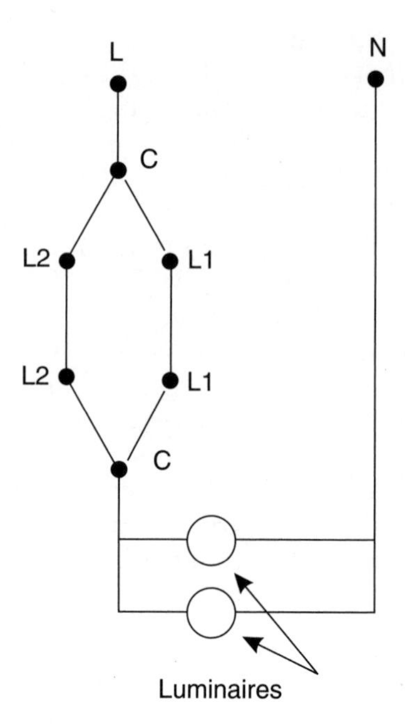

Figure 15.6 *Two-way switch*

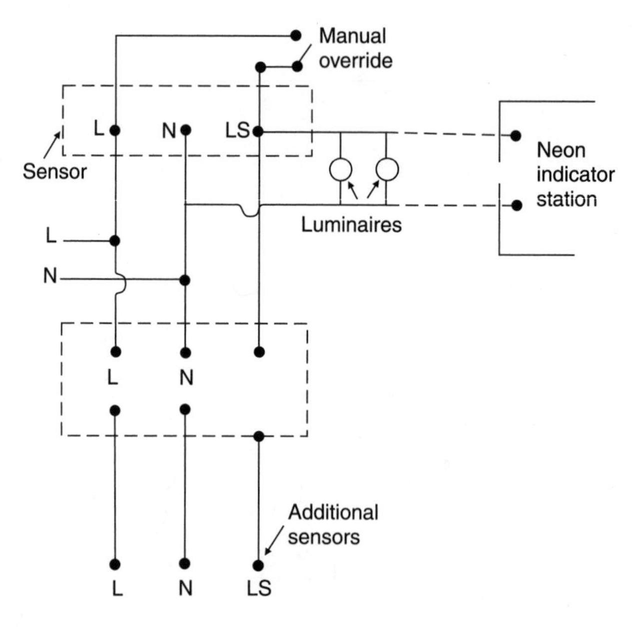

Figure 15.7 *Neon across lighting load*

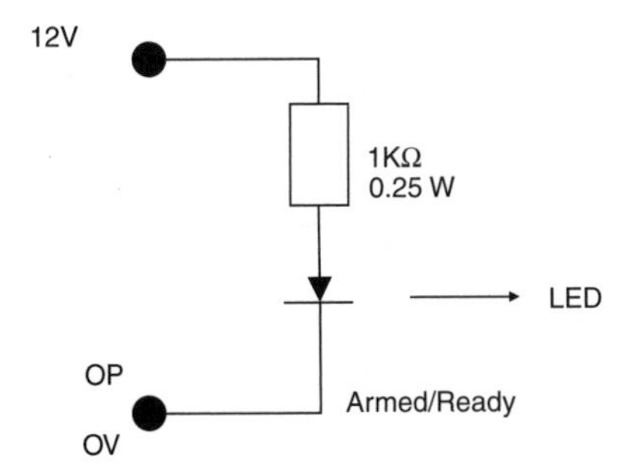

Figure 15.8 *LED indication*

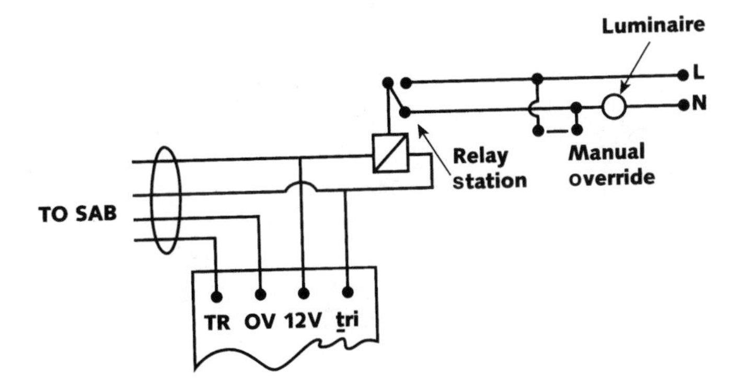

Figure 15.9 *Luminaires follow an alarm output*

with an intruder alarm system. Figure 15.9 illustrates how this can be achieved. Manual overrides can then be added so that the lighting is not solely for use in the event of an alarm condition.

Many security systems also feature programmable high current transistorized output connections of different states and with a choice

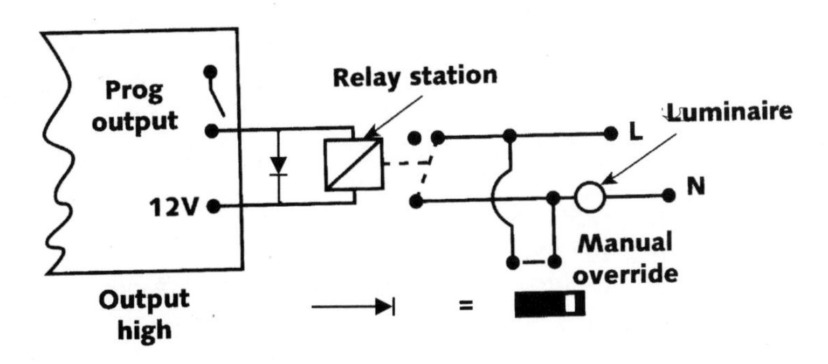

Figure 15.10 *Programmable output*

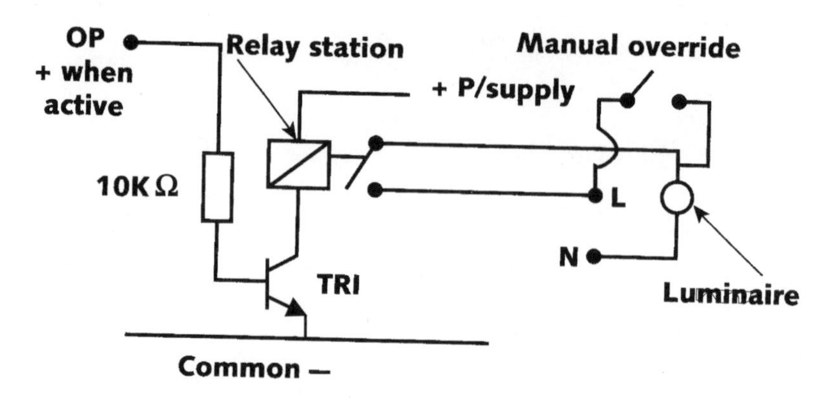

Figure 15.11 *System output current limited*

of types. One well-established output type can be programmed to follow the buzzer operation or exit/entry time through an intruder alarm system or the timed passage through an access control system. This output may be selected automatically or by an additional keypress. Such an output can energize a relay to activate lighting through this time period to illuminate a designated route. If the system output is current limited an NPN transistor to switch the relay coil can be added (see Figures 15.10 and 15.11).

It becomes quite apparent that security lighting systems can be customized in their wiring technique to suit individual needs. Although a mains sensor head may not have additional contacts to bring in optional equipment such as buzzers and warning devices it is always possible to add components to a system to expand its range of functions. Figure 15.12 shows a mains PIR sensor with a buzzer and a separate photocell to show how the lighting and buzzer can be switched for day and night operation.

By reference to previous figures it is possible to devise a means of adding indicators to show the active state of the circuits.

The technique that appears in Figure 15.12 shows how it is possible to configure a simple system to allow a mains automatic sensor to drive a buzzer or sounder at the times when the photocell opens the circuit (LS live switched wire) because of the availability of natural light. At those periods of time the luminaire would be prevented from energizing. In this case the photocell has volt-free contacts. A switch in series with the buzzer can disable the audibility of the system so that for instance the buzzer can sound during the day but the luminaires operate at night.

In the event that an extra low voltage system is being used to power a particular automatic sensor the buzzer could be driven via a relay

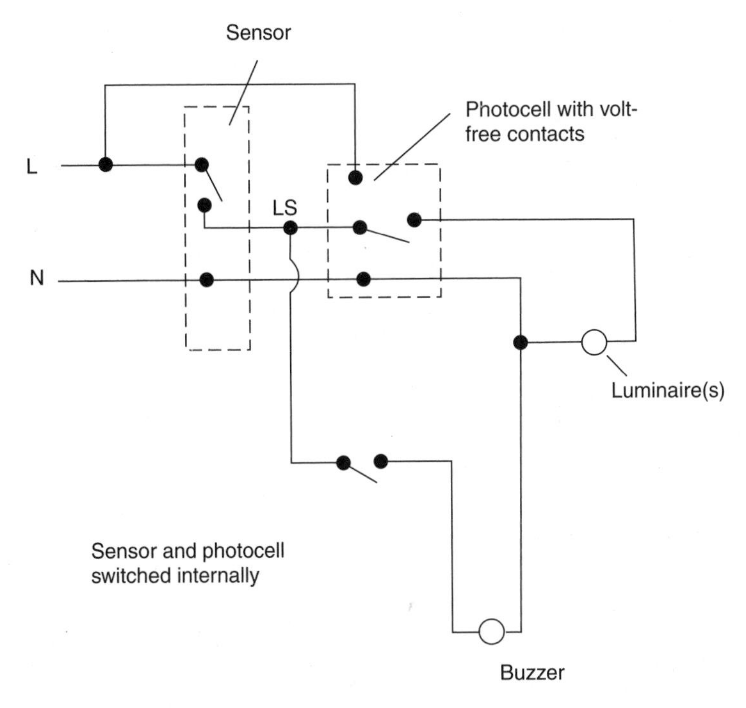

Figure 15.12 *Sensor with buzzer and separate photocell*

and ultimately by the power supply that is being employed to energize the detector. Alternatively if a mains buzzer was not wanted or available it could be replaced by a relay with a 240 volt coil and a low voltage buzzer could then be driven by the power supply.

The system is therefore governed in practical terms by the selection of the automatic detector and the supply voltage that is used to drive it. As previously stated it is always possible to add indicators such as an LED or a neon indicator to show the state of the outputs.

An option to using mains sensors and customizing these to suit an application or special needs is to select a true touch control lighting system with extra low voltage high quality sensors. These are practical with clearly defined wiring techniques and have individually managed zones. They therefore provide versatility in the selection of light on times and manual lighting overrides with selectable audible tones for each particular zone. Indicators on the touch control monitor activations from the sensors and identify the accurate location of the detection. Additional touch controls can be added to enhance the system and provide control from more than one location. The lighting is switched by separate mains expansion units that can be placed in any location within easy access of a mains supply and convenient to all cable runs. These connect to the main controller with extra low voltage cable. This allows for a neat installation to what could have been a difficult location. These systems are ideally suited for all domestic or industrial applications especially where additional touch controls are required. Expansion units enable an even greater control of lighting to be achieved.

Demand lighting is indeed available to suit all budgets. Examples have shown a number of wiring configurations to illustrate the many different forms that it can take in the practical sense to include two-way switches. If automatic detection is not wanted but an area is to be illuminated well in advance of a person walking into such a sector, such as a dark passageway, this lighting can be manually selected by a two-way switch. This can then be switched off manually at the end of the passageway. Neon indicators can always be added to show that lights have actually been turned off and these can be at various points or be amalgamated with the switches. If manual switches are being used outdoors weatherproof versions are available. These often have neons included in their construction to better identify the installed location or position of the switch.

Although demand lighting can be amalgamated with many other features and customized to suit any particular requirement it is possible to add other components and warning devices to extended period lighting. Therefore even in areas that are kept illuminated for long time scales sensors can be added in order that outputs can be brought into effect if movement by persons or cars is detected.

The essential aspects of lighting that is intended to be used over long periods of time is broached in the following chapter.

16 Extended period lighting

This type of lighting is referred to as extended period lighting because it is intended to stay on for reasonably long periods of time such as from dusk to dawn. It is not intended to be switched on by automatic sensors that detect movement of a person or a vehicle and then only keep the lighting energized for a short term.

Because lighting for extended periods covers a different type of application to that of luminaires selected on demand it also uses different light forms as the luminaires that are used are more governed by economic restraints on current consumption. It must have low running costs and provide a light form that gives a reassuring environment. The lamps are therefore of low energy consumption with different versions being specified for applications where aesthetics are important particularly in the domestic sector. The lighting may take on a slightly separate guise for mainstream commercial and industrial lighting although traditional fluorescent tubes in a protected form will be found in some applications.

One of the true advantages of lighting that is left on for long periods of time is that intruders cannot feel comfortable when entering an area within its presence. Therefore they seek to extinguish the light form but this brings attention to the area that is generally perceived as being well lit. Intruders are therefore faced with the dilemma of entering an area already illuminated and risking being seen or attempting to turn off the lighting but wondering what effect this could have in attracting attention to the sector.

It is not normal practice to select extended period lighting for security purposes by manual switches alone but these tend to be added to an automatic technique of bringing on the luminaires. Therefore the manual switches override the automatic function. The automatic function is generally selected by timers and/or photocells that can be separate or integrated with the luminaire.

There is a good range of timers and photocells available with options on their use. We tend to associate the term photocell with a standalone device that is capable of handling luminaires directly although these are still available in a range of different ratings. Certain devices will therefore be unable to switch heavy loads directly and must be used with contactors as interfaces. The reader will also come across the terms LSD and LDR. The LSD or photodiode is a light sensitive diode that is used to activate a circuit in response to light. It is an electronic component and needs additional components in its circuitry to function. This is also the case for the LDR or light dependent resistor which is similar to the LSD but is capable of handling greater currents. It is used as a switching device in such applications as street lighting.

There is now a growing trend to use low energy lamps and to recognize the need for energy management and replacement lamps that can be retrofitted into traditional installations. This is considered further in Chapter 18.

Therefore in cases of installing lighting intended to be energized for extended periods the installer will be involved with discharge lighting. This relies on the ionization of a gas to produce light. In these devices high voltages are present so special precautions for the different versions and as detailed by the manufacturer and outlined in the IEE Wiring Regulations are to be upheld. The associated control gear of discharge lamps is highly inductive and there is a need to consider any inrush currents alongside the selection of the cables and the circuit protection. The value of the design current of the equipment will be offered by the manufacturer of the equipment. There should also be a recommendation as to the optional means of isolation from the supply at the origin of the installation to cut off all voltage as may be necessary to prevent danger.

In all systems the installation is to have a means of isolation, a means of overcurrent protection and earth leakage protection. This applies whatever the size or type of installation. Industrial applications will only differ from domestic installations in the size and type of equipment used.

For extremely heavy loads, switch fuses are replaced by circuit breakers and there will be overhead busbar trunking systems. To satisfy the IEE Wiring Regulations the circuits must be designed and the design data made available. There are a number of steps to take in this process but these are the same as for any electrical installation. Important is the means of overcurrent protection which should comprise either a fuse or circuit breaker inserted in each phase conductor of the supply.

Fluorescent tube systems have tended to be classed outside of true discharge lighting because of their use in the domestic environment and their presence is well established. However it remains to say that in fact their characteristics are different to those of filament lighting. The essential circuit wiring requirements of fluorescent lighting are as those for filament lighting but it cannot be adequate to convert the nominal lamp watts to the exact equivalent current. This is because, as is the case with all discharge lamps, additional current is drawn owing to losses in the control gear, harmonics and power factor. This does, however, depend on the circuit used. The actual current should be made available by the manufacturers in their data sheets. If this is not available it is normal with discharge lighting circuits in voltamperes to be taken as the rated lamp watts to be then multiplied by a factor of 1.8. This figure is intended to take into account control gear losses and harmonics assuming a power factor of not less than 0.85 lagging.

In order to appreciate typical electrical data discharge circuits we can refer to Table 16.1. This illustrates the relationship between values and represents the figures that a manufacturer would generally quote for standard units.

When choosing fluorescent control gear it is to be noted that there are three electrical circuits in common use. These are namely switch start, electronic start and high frequency starting techniques. They are not to be mixed on the same circuit or damage can occur on high frequency ballasts.

Table 16.1 *Electrical data discharge circuits (data is rounded for clarity)*

Lamp type	*Total circuit (watts)*	*Ballast loss (watts)*	*Start current (A)*	*Main running current (A)*	*Power factor (lagging)*	*Capacitor value (microfarad)*
SON						
50 W	62	12	0.45	0.29	0.89	8
70 W	86	16	0.60	0.39	0.92	10
150 W	172	22	1.20	0.83	0.86	20
250 W	276	26	2.30	1.35	0.85	30
400 W	431	31	3.60	2.10	0.86	40
Metal halide						
70 W	81	11	1.00	0.37	0.91	10
150 W	170	20	1.20	1.10	0.91	16
250 W	276	26	2.30	1.35	0.85	30
400 W	431	31	3.60	2.10	0.86	40
Mercury						
50 W	62	12	0.40	0.27	0.95	8
80 W	94	14	0.65	0.44	0.89	8
125 W	141	16	1.10	0.67	0.88	10
250 W	275	25	2.20	1.32	0.87	16
400 W	435	35	3.90	2.07	0.88	25
SOX						
18 W	26	8	0.16	0.13	0.83	5
35 W	52	17	0.27	0.26	0.83	6

Switch start is the simple technique of using a magnetic copper/iron ballast, a capacitor and a glow start canister. Electronic start adopts an electronic starting device instead of a conventional starter switch and gives flicker-free starting. It warms the tube before the starting operation so this leads to extended tube life. Electronic starters also automatically switch off failed tubes to prevent flicker/flashing.

High frequency new generation circuits run at almost unity power factor reducing the VA load and also ensure flicker-free soft starting. In addition they achieve extended tube life and automatic closedown of failed lamps. They are completely silent in operation. Digital high frequency regulating also enables dimming to be achieved so it is used with energy management systems. This digital regulating provides extremely good control of light output to be accomplished.

There are a number of recommendations provided by discharge lighting manufacturers for the form of circuit protection to be employed on applications using their products. These include the use of miniature circuit breakers (MCB) devices. These are an alternative to fuses and can be mounted in a distribution board or consumer unit in much the same way as fuses. These are switches that are designed to open automatically when the current passing through them exceeds a preset value. These devices should have time delay tripping characteristics in order that the operating time is controlled by the magnitude of the overcurrent. Thus the circuit breakers will not be affected by transient overloads and the switch-on surge of the discharge lamps. They will only operate and 'trip out' if the overload is present long enough to constitute a hazard to the circuit that it protects.

The MCB will therefore give overload protection but tolerate the normal higher starting current of the lamps.

MCBs have an advantage over fuses in that once having operated they can be reset. They are also very accurate in detecting the trip current and provide a high degree of discrimination. Nevertheless despite the ongoing use of MCBs in many traditional industrial installations high rupturing capacity (HRC) fuses will still be found. These may also be seen protecting motor circuits since they can discriminate between a starting surge and an overload. In practice they are fast acting and consist of a porcelain body filled with silica. They have a silver element and lug type end caps. An indicating element is used to show that the fuse has ruptured.

All protective devices provide a different level of protection such as rewireable fuses being slower to operate and being less accurate than MCBs. In order that all such devices can be classified we need to know their circuit breaking and fusing performance. This is achieved by the fusing factor as:

Fusing factor = fusing current/current rating

This equation covers the fusing current as the minimum current that will cause the fuse to blow.

The current rating is the maximum current which the fuse can sustain without it blowing.

From this we can say that a 5 A fuse which only blows when 9 A flow has a fusing factor of 9/5 = 1.8.

HRC fuses have a fusing factor up to 1.5 maximum and circuit breakers are designed to operate at no more than 1.5 times their rating.

If extra ways are to be added to HRC fuses Table 16.2 can be used as a guide as to the ampere fusing details for discharge circuits. This is based on 240 V 50 Hz.

Many manufacturers of discharge lighting also advocate the use of surgeproof inert ELCB or RCD devices to be used alongside the appropriate protective devices to detect leakage current but with a 100 mA breaking capacity in preference to 30 mA so that nuisance tripping is avoided.

In order to achieve a respectable degree of security for the installation there are further needs beyond those of circuit protection to consider. In this sense we need to appreciate a want for protected wiring, standby power, access to luminaires and cables and separation from other circuits so that the lighting cannot be readily disconnected. There is an equal need to address the problems of circuit faults developing and how this can stop the operation of other luminaires in the system so circuits must be developed to compensate.

On the understanding that the lighting is to be used for long periods of time there is a further need to address light pollution and the effect of external lighting on properties local to the area being illuminated.

Once the design data has been made available and drawings produced the circuit diagrams can be established. In the event that the luminaire has the timer or photocell integral in the housing the installation is relatively straightforward because the switched connections of the assembly are prewired and extra terminals are generally also provided for the manual switch function.

Table 16.2 *Guide to fusing details for discharge circuits (A)*

No. lamps per circuit	*1*	*2*	*3*	*4*	*5*	*6*
70 W	4	4	4	6	6	10
150 W	4	6	10	10	16	16
250 W	10	16	16	20	20	20
400 W	16	20	20	20	–	–

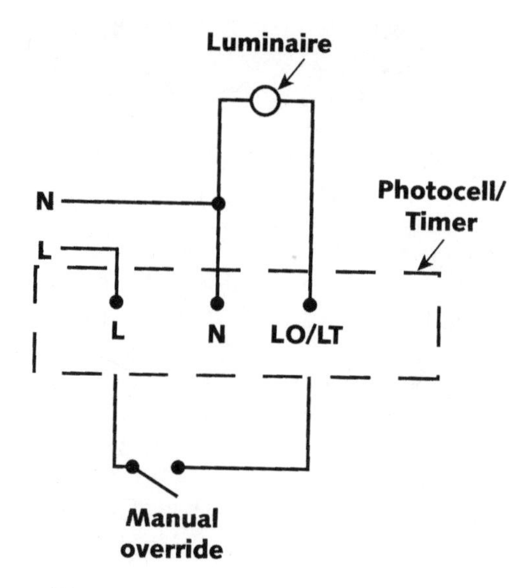

Figure 16.1 *Photocell/timer control over luminaire with manual override*

The connections for a system with separate luminaires and photocells or timers are shown in Figure 16.1.

It may be noted that the wiring of the timer and photocell circuits is different to that of a traditional two-way switch although there are two switched functions. The reason is because both the timer and photocell reset automatically and in practice the manual switch is only wired across these in a parallel circuit. The two-way switch configuration was shown in Figure 15.6. A neon can always be wired across the same terminals as the load so that it is possible to determine from a remote location the state of the luminaires and at what times they are illuminated.

One of the major benefits of lighting that is illuminated for long periods of time and that has been well designed and is energy conscious is the fact that it is always overt. Therefore it can be seen to be fault free and in regular use. However, lighting that is only called up automatically and for short periods of time is not subject to the same degree of employment so it is unable to offer the same level of reassurance.

With all systems there still remains a need for a planned maintenance procedure to be put into force to satisfy the IEE Wiring Regulations and any particular needs that the system requires to satisfy its duty. This is to include cleaning and lamp replacement of all luminaires. It is important to take this into account because although security lighting is also used in internal applications it has many uses for exterior lighting and the access to lamps in these applications is more difficult.

Exterior lighting can always be subject to abuse by vandals and intruders because it is difficult to protect the lamps with other security technologies in the same way that it can be accomplished with internal components. The principal reasons for all lighting schemes but in particular exterior lighting are deterring crime against people and property, providing safety and making the night scene more attractive. However, to meet these aims there is a need to address the problems of abuse. There is a role to play for robust bulkheads, heavy-duty car park wall mount luminaires, steel housed ground mounted bollards and spheres on brackets at high elevation points. The only serious means of making lamps difficult to gain access to is by employing floodlights or floodlight columns and further improving their deterrent value by keeping them energized for long time spans.

Floodlights for discharge lamps in a polycarbonate enclosure are able to withstand abuse from stones and other projectiles. They can be mounted up to 6 metres high on the side of a building or be pole clamped to ensure even higher durability. High output devices can be installed at progressively greater heights.

Visors may be hinged to the body to enable easier installation and maintenance with the gear mounted on a removable tray. These can be specified with a variety of lamps.

The beam form of the lamp or its orientation will be specified by the manufacturer. These will range from narrow beams through to medium (symmetrical and asymmetrical) and extend to wide beam patterns. They will vary enormously depending on the manufacturer of the goods and the intended light spread. However, the area that can effectively be covered is easily established by relating the beam pattern against an elevation drawing of the area to be illuminated.

Inclinometers can be incorporated in the floodlight assembly to make adjustment and aiming easy and the data will give guidance on the spacing between the units as related to the mounting height.

Floodlight columns are intended for root mounting by placing them into an excavated hole in the ground and supporting them in a perpendicular position. The duct for the cable is placed into a cable entry slot and the hole is backfilled with concrete to the amount suitable for subsoil conditions. Alternatively the column can be placed into buried vertical pipes supported by a mass of concrete. The column is then grouted to the pipe gap to give a firm fix. Brackets are secured to the column top to hold the luminaire and the cables run within the column so that they are fully protected.

The columns are available in a multitude of forms with floodlights or robust spheres matched to them. They can have tamper resistant locks and attachments with the luminaires held at elevated positions in the order of 10 metres in height. Although many columns reach heights of 10 metres it is possible to use purpose floodlights for sports stadia, car parks, ports, airports, railways, industrial yards and motorway intersections. These will have columns from a height of 16 metres up to 25 metres and tend to include 2000 W metal halide or 1000 W SON lamps. The surveyor is therefore given an opportunity to select a column to hold floodlights or spheres which is difficult to access by intruders or to damage by projectiles. These columns themselves will vary considerably in their weight, area projected towards the wind and nominal heights. If multiple luminaires are used at the column tops an additional level of security is reached in compensating for bulb failure and downtime. A number of luminaires or floodlights sited at strategic positions at the top of a column can provide an area of cover of 360° and compensate for each other in the event that a bulb develops a fault.

If it is possible that shadows can be cast by the close proximity of walls, flush lamps can be added as surface mounted units in the building structure as shown in Figure 16.2. These robust anti-vandal lamps were considered in Chapter 12 in our observations dealing with the type of luminaire and can provide a function of revealment lighting. Columns that are complete with floodlights or spheres are often employed in systems that are amalgamated with CCTV as detailed in the following chapter. These may also be supported with additional light forms such as the aforementioned flush/concealed low level

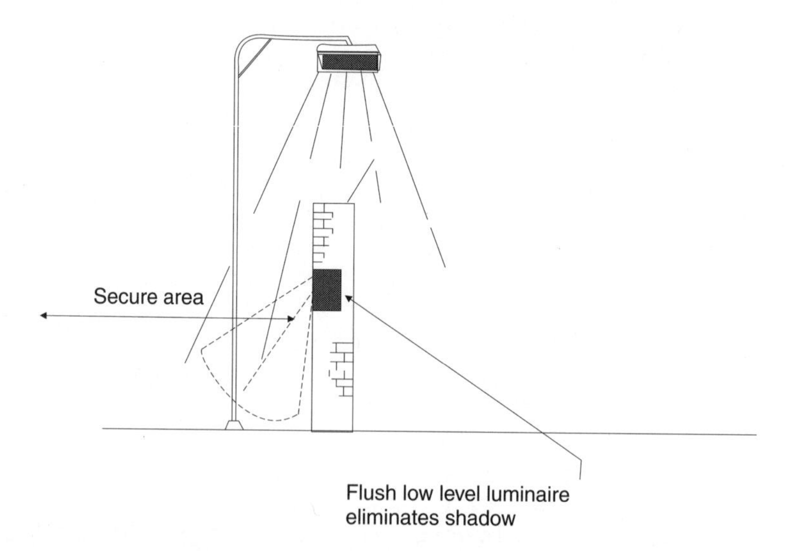

Figure 16.2 *Revealment lighting*

lamps. In the event that additional lighting is needed for high security risk applications and to portray objects such as vulnerable storage tanks, surface mounted luminaires can be mounted on these structures to double up on the illumination and to stop shadows being formed at their bases.

As a conclusion to this chapter we may add that maintained emergency lighting can also double up to give a good security lighting feature by enabling constant illumination to be achieved. This can be used to great effect for internal applications in commercial premises and in particular those in which cameras are used to monitor the area at all material times. Permanent illumination is an excellent deterrent in these circumstances and the deterrent value can be further heightened by mounting a slave monitor on display to illustrate to potential intruders that observation networks are in constant use.

17 CCTV and observation lighting

The use of CCTV is now widespread and no longer confined to the major commercial and industrial areas, and is becoming more accepted and of increased significance in all sectors including the domestic environment. Indeed the use of visual monitoring systems is being extended with the introduction of verification of personnel and events into the mainstream security market. This increased activity has made prices of components more attractive and catered for the more cost conscious of applications. A progression on this is a greater need for artificial lighting during the time when natural light is not available as 24 hour surveillance is something which many customers seek.

There are many considerations when designing a system for all hours operation and one important effect on image quality is illumination of the area being monitored. When surveying for external applications it will be appreciated that there is a stark contrast between levels of night and day illumination. It is clear that during daylight conditions it is easy to capture effective images from a CCTV system when the correct components are specified. However, during the hours of darkness artificial lighting must be added to the system in order to produce usable recordings of events and to observe activities as they take place. Essentially, without light there can be no picture. There are therefore three main subjects to address, namely illumination, cameras and lenses. From this there must be a combination that enables all the specified products to work together as a group and to maximize the benefits offered by each other.

Government figures indicate that crime is on the increase and it is a fact that a large percentage of crime actually takes place during the night. For this reason the CCTV industry must focus its attention on achieving effective 24 hour monitoring. In theory this means night-time footage must be as effective as that achieved during the hours of natural light. It is important that the cameras can continue to record from dusk to dawn and maintain their capacity to provide images with respectable contrast and good resolution. The means of achieving this vary and in some instances there is a need for infrared lamps which emit a light form invisible to the human eye so can not cause any light pollution. In other cases the luminaires employed may be similar to those used in traditional extended period lighting duties for providing normal reassurance and security lighting. In theoretical terms there must be light reflecting and refracting off the surfaces of an object if it is to be seen. It is a frequency of above 700 nm which represents the boundary of visible light.

The responsibility of ensuring that the lighting enables the observation system to fulfil its role lies with the installer. There is also a need to prevent light pollution as the lighting is to be restricted to the points that the cameras actually view in the same way that the cameras must be directed to the same clearly defined area. Before the selection of the luminaires and their siting is determined it is necessary to determine exactly what is the aim of the observation system. The considerations of the installation will involve not only the positioning of the luminaires but also the distribution of the light emanating in the area and the exact nature of all of the lighting sources.

When designing a system for CCTV monitoring for 24 hour surveillance, particular attention must be paid to the night-time performance of the camera, lens and illumination. In so far as the cameras are concerned certain types are better suited to working at night than others. All cameras are not the same and some are better suited to providing effective cover at night. Nevertheless all modern CCD cameras do offer some degree of infrared response, which makes them suitable for use with low power IR sources such as LEDs. We are at the stage at which dual mode cameras (day–night, dual technology) are used to provide the best compromise for 24 hour surveillance using colour by day and mono/IR sensitive by night. These generally have small moving filters that are moved over the CCD sensor for daytime/colour operation and then moved out of place during the night-time for monochrome operation to maximize low light sensitivity. Other camera designs incorporate specialized filters that are non-moving and provide both good colour performance and IR sensitivity. Traditionally the industry has used colour cameras to identify and track targets but for observing in poor lighting conditions black and white/monochrome has been used as colour cameras need more light to operate than monochrome.

In most cases it is probable that some lighting already exists. It is even possible that this lighting is adequate for some of the cameras that are to be used so the CCTV system can be surveyed and installed in certain ways to integrate with the existing lighting. It is important to recognize this philosophy because the client will naturally show a measure of reluctance to pay for additional lighting to support the observation techniques when a certain level of illumination in the local area is already in use. This is because it may have stood the test of time in its efficiency for amenity purposes. It is possible that only a measure of secondary luminaires needs to be added to obtain a good level of illumination but the position of cameras can be selected to take into account the existing lighting network. This is because black and white cameras are able to utilize the infrared in the environment which is to their advantage in darker areas.

The key elements to consider when selecting a camera are its sensitivity and low light performance, its signal to noise ratio and the spectral response which is the ability of it to see IR. There is also the night-time performance of the lens to be considered and its light-gathering capability. This must be maximized because daylight and infrared light have different focal lengths because the different wavelengths of light pass through the lens differently and so do not focus at the same point. This causes a shift in focus between daytime and IR operation. The extent of the focus shift depends on a number of factors including the quality of the lens and the wavelength of the IR filter although we are seeing the development of lenses with zero focus shift between daytime and IR performance. The key to successful night-time schemes is to sufficient light, the right quality of light and proper control over it.

When there is no lighting any illumination and cameras must be seen as an integrated system because it is incorrect to determine the position of the cameras and then at a later stage select the luminaires. The security lighting is committed to perform its duty as such and also to enhance the observation requirements of the CCTV technique. Nevertheless when the two different techniques are amalgamated the performance of each as separate identities is extended.

We are mainly interested in the performance of the security lighting as opposed to that of the observation technique but it is clear that it plays an essential role in allowing the cameras to perform adequately.

Having come to accept that there are many different types of lighting and these range from ultraviolet through to infrared we must know that the human eye is not capable of seeing all of the available light forms. In general terms light that is classified outside of the 350 to 750 nanometres range is not visible and this includes both ultraviolet and infrared lighting forms. However, infrared lamps can be confidently used as a source of illumination in CCTV applications and are most suitable for covert surveillance.

The different types of luminaires will give various results so it is normal to decide exactly what are the needs of the observation system and to give some thought to the equipment that is being used.

Colour rendering properties must be attended to when selecting a luminaire for use with colour cameras because the quality of the final colour on the monitor will be governed by the colour rendition offered by the lamps. It will be found that the most superior light form for colour cameras is white light but it is not always possible to employ this form as it is expensive and not always convenient. In addition it can cause light pollution in certain instances so it is important to consider the background and where it is to be used as it may not be the best option if it is to be used for long periods of time. The surroundings are to be checked together with the local roads before using white light.

If the intention of the security lighting is to catch intruders rather than deter them it is worth considering the use of infrared lamps because these are ideal in covert operations. Infrared lamps can always be used in CCCTV applications without having to install additional luminaires. They utilize a light form that is not visible to the human eye but can still be detected by particular cameras. The normal technique is to install the infrared lighting at the camera head and control it by either telemetry or photocell. Since this infrared light cannot be seen by intruders they have no means of recognizing the areas that are being observed by the cameras and therefore potential offenders become intimidated by its use. Although certain infrared lamps, particularly those that operate at frequencies in the order of 750 nanometres, emit a slight red glow at the face of the lamp, which may be seen by the human eye, this does not in any way reveal the actual area being viewed. It will be noted that infrared lamps that operate at even higher frequencies do not emit any detectable red glow. In general these cameras operate between 715 nanometres and 950 nanometres. The cameras that operate in the order of 715 nanometres produce the greatest level of red glow whereas those that function at wavelengths around 830 nanometres greatly reduce the visible glow. A 950 nanometre lamp is totally black and ideal for covert use where no indication of night-time security is a requirement although a lamp of this nature needs a highly sensitive night-time camera.

In all cases these infrared lamps are intended for use in applications in which the intention is to catch offenders as opposed to deterring them. The other role that this form of lamp can play is in those instances where there is a want not to install additional luminaires or to create a greater lighting effect than is already in existence. These lamps therefore are invaluable in cases where light pollution could be a problem.

It is of vital importance when choosing infrared illumination to consult the manufacturers of the equipment since all infrared lamps must be given a match to the camera that they are ultimately to be used with. All units do not necessarily work with each other although certain cameras are promoted for operation with infrared lighting. These tend to be defined as high level infrared sensitivity cameras.

It may be known that monochrome cameras use some of the infrared light available in their operation so these are better used with infrared lamps because colour cameras are not so infrared sensitive. Day–night cameras can actually be used to capture colour images during daylight conditions and then adjust to monochrome when the light level drops below a certain threshold. A photocell can be used as an option with monochrome cameras to switch on the infrared lamps once a given lux level is approached.

A part of the surveying and the maintenance procedures must take into account the illumination range and the life expectancy of the lamps. Although the illumination range depends on the normal ambient light levels and the weather conditions it will be found that we can still quote maximum distances for the lamps. These are 15 metres for a 30 W bulb, 30 metres for a 50 W bulb and up to 200 metres for a 500 W bulb. Nevertheless the ranges will vary and can only be used as a guide.

At the time of the survey it is of equal importance to obtain the life expectancy of the bulb so that this can be budgeted for during the maintenance or for normal stocking purposes. IR lamps are often found used in pairs especially with pan and tilt cameras. In these cases one lamp illuminates a wide-angle view to enable general surveillance and the second lamp is used to provide illumination for the zoom view. They are found mounted on either side of the camera and projected in the direction of the camera view and cannot be directed towards the camera as the result will be a 'white-out' of the camera image.

The benefits of infrared include:

- Although infrared is invisible to the human eye monochrome cameras can see it.
- There can be no light pollution.
- The lamps are discreet or covert and easily blend in with the surrounding environment.
- Intruders are unaware of the surveillance.
- There is no glare produced to irritate pedestrians or vehicles.
- Long ranges beyond normal lighting systems are achieved.
- The installation and capital costs are low.
- The running and maintenance costs are cost effective.

In general the main choices with infrared illumination are the power of the unit, the field of view and the wavelength of light being utilized. Filters of 730 nm are brighter in appearance to the observer than 830 nm or 950 nm filters but they do provide more usable infrared radiation for the camera. In addition the redder appearance glow of 730 nm filters does act as a deterrent. The field of view of the lamp must also be matched to the camera. It is necessary to provide even illumination to allow the camera to work to its dynamic range. For shadow-free pictures the lighting must always point in the same direction as the camera. For covert surveillance in dark conditions infrared illumination is the only solution. However, there is a need for adequate light energy at the scene and to match it to the sensitivity of the camera if it is to have an appropriate signal to noise ratio.

It becomes understood that although there are some advantages in using infrared lamps there are other applications in which it is more desirable to use extended period lighting or even automatic pulsed security lighting for short time periods.

For extended period lighting low pressure sodium lamps are not recommended as their colour definition is poor so that the colours on a CCTV monitor cannot be easily distinguished. This leads to most objects appearing as grey. They only have a limited role to play if there is no requirement for clear colour images or fine detail display. They can be found in use in systems with monochrome cameras if colour rendering is not vital but can suffer if the images are needed on video for court purposes because they will not be able to provide clear images in order to counter legal challenges to suspects. However, low pressure sodium lamps do benefit from a high luminous efficacy together with low running costs and can be replaced by low pressure sodium plus lamps which have a greater life expectancy.

High pressure sodium lamps are a better alternative to the low pressure type as their golden white light enables colours to be more easily distinguished. They also have a high luminous efficacy and pleasing colour source so can be confidently employed as amenity lighting although they are not as economical as the low pressure variant. A long life span is offered by high pressure sodium lamps and they are not adversely affected by low temperature working.

Mercury discharge lamps exhibit a cool white light and also provide good quality colour rendering but do not have the efficiency of low pressure sodium lighting and also suffer from a reduced lamp life.

Metal halide lamps are also classified as high pressure discharge luminaires and produce a clear white light that allows good colour rendering particularly in floodlight form. They have a reasonable lamp life expectancy. Fluorescent lighting is extremely economical to run and with its high efficacy it also provides a good white light source. Tubular fluorescent lamps produce a light form between cool and warm white so their colour rendition can be largely selected. Compact fluorescent luminaires are becoming increasingly popular as they produce greater levels of light at a lower running cost with long lamp life.

Automatic lighting that is pulsed for short periods tends to use tungsten halogen floodlights and although they also provide good colour rendering they are expensive to run and can suffer from short lamp life.

It is important to remember that white light always enables colours to be clearly defined so mercury discharge, metal halide and fluorescent are always capable of providing good images on CCTV monitors. Low pressure sodium although economical is not suitable for use with colour cameras but it has a limited role to play with monochrome systems. High pressure sodium is a good choice because it is capable of good colour rendering and can be used for extended periods of time because of its relatively low energy consumption. Tungsten halogen lamps can have a role to play in CCTV lighting networks as they offer good colour discrimination because of their brilliant white light source but they are not economical when used over long periods of time.

If covert operation and observation are important infrared lamps can be specified alongside monochrome cameras. It is important to remember that infrared lamps cannot work with colour cameras but dual mode cameras are a good option which will work with infrared lamps when the camera is in monochrome mode. The distance or the range over which the camera will be able to operate in the dark is very much governed by the sensitivity and spectral response of the camera and the lens combination. Manufacturers provide charts to indicate the normal performance expected.

Although it can be said that infrared lighting is discreet and covers a wide range of frequencies it is also highly reflective and details can be lost on light surfaces. It is normal to match the camera to the light and it may be beneficial to use a number of lamps to provide an overall scene. For instance a wide-angle flood lens can provide a wide picture and a narrow spot lens can give an accurate beam to penetrate dark areas.

As an overview of the chapter, cameras for 24 hour monitoring have defined specifications but for image efficiency cameras require compatibility with the visual scene. This is with respect to both the correlated colour temperature and spectral sensitivity. The correlated colour temperature of the camera must be matched to that of the scene so this will preclude some light sources such as low pressure sodium. The spectral sensitivities of CCTV cameras do not normally have the same characteristics as the human visual system so the light source must take this into account.

For true covert operations infrared lamps can be used but these must be matched to the camera and scene.

While all lighting provides illumination it does not always do so at the same frequency but must operate within a band. This frequency band is as shown in Figure 17.1.

Colour cameras are designed to work efficiently across the full white light range so can be used in areas with fluorescent lighting tungsten halogen lamps, standard white light bulbs, white lighting discharge systems and natural daylight. Sodium vapour lamps cause an imbalance in the illumination because they do not cover the blue end of the spectrum and this has an effect on the colour reproduction.

In conclusion the lighting in any 24 hour CCTV system is a priority and must also be balanced against the risk of failure or vandalism if the observation network is not to be compromised.

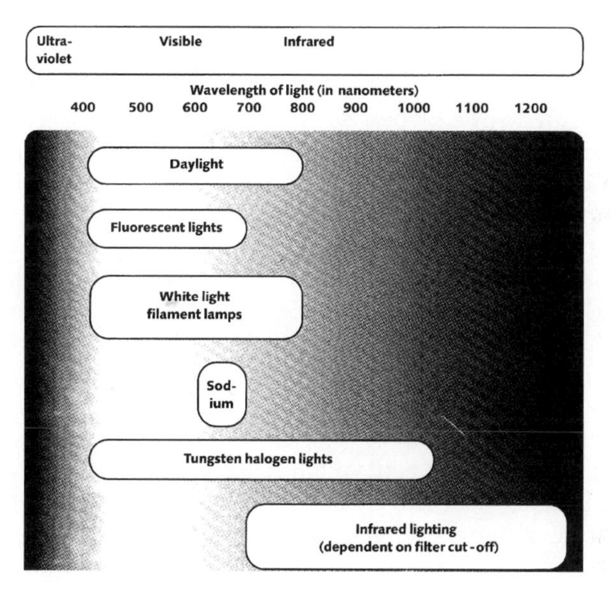

Figure 17.1 *Lighting frequency*

18 Energy management

The efficiency and management of lighting is becoming a priority in the commissioning of new buildings and in the upgrading of existing systems. Indeed the subject of energy management is expected to become one of the most important considerations within the Building Regulation documents and will have a tremendous impact on the way that the construction industry looks at energy. It is apparent that serious measures must now be taken to reduce energy usage and wastage. This will have an impact on security lighting and the way that it is applied. Lighting experts will show an increasing urge to work alongside electrical contractors and installers in order to help them increase their business opportunities by identifying the roles and applications in which energy efficient lighting should be installed. Electrical contractors are being more educated in the use of design lighting that is effective and energy efficient.

The needs are therefore to equip personnel to:

- Recognize any inefficient installations.
- Come to appreciate exactly what are the environmental, cost and associated benefits of energy efficient lighting schemes.
- Estimate energy cost savings and calculate the payback period.
- Recognize the situations in which expert and specialist knowledge is needed in the design of management systems.
- Think in terms of increasing business but with an acceptance of having also to save the environment.

At certain points in time it was said that the brighter the lighting of any system proved an advantage. However, we are now seeing a trend to take us away from large floodlights illuminating the night sky with a strong white glare as exterior lighting is becoming much more focused on the minimum lux levels required. We are also seeing a move towards directional beams.

The lighting industry wants to remove itself from a proliferation of public and private external lighting schemes to counter the light pollution problem and to become more energy and cost conscious in its make-up. There must be a mechanism to tackle the problem of countless floodlights, uplighters, spotlights, decorative installations and an array of security lighting forms that are badly installed and specified, create light pollution and use high energy levels.

Lighting pollution is now at the forefront of debates for two main reasons:

- Light pollution spoils the natural effect of the night skies.
- The greater the light pollution the greater the power consumption.

Unfortunately there must be a certain degree of light pollution in order to satisfy safety and security applications. Equally there must always be a desire to have purely decorative lighting installations so the answer lies in a compromise. Systems must be designed with a degree of thought given to the avoidance of light pollution and energy wastage. External lighting must provide minimal light pollution, a safe environment and an attractive feature. For attractive features we will see a greater use of fibre optic solutions with colour changing effects and for bay lighting the lanterns are to be engineered to direct the illumination downwards. Bollards or recessed ground luminaires can be set into walkways so there is no spill into the night sky. Intelligently designed schemes can ensure that lighting is only reflected in a downwards direction so that pedestrians are better guided, the lighting is of a pleasing effect and there is little light overspill.

It is therefore recognized within the lighting industry that there is a need to raise standards in all aspects associated with light and lighting – in particular when it comes to energy management and light pollution. There is a need to define and harness the pleasures of lighting but at the same time promote the benefits of well-designed, energy efficient schemes among the public at large. There must also be miniaturization and increased lamp life. Energy management must therefore be a part of security lighting.

Demand switched lighting has always been regarded as cost saving because such lighting is only available when it is required.

Low energy extended period lighting can also exploit an opportunity of replacing lighting that is wasting energy. In the first instance we can look at a number of ways in which we can recognize waste and exploit the opportunity in order to gain a benefit. These all have security lighting involvement.

By improving energy management we can make greater use of security lighting so it becomes apparent that a part of security lighting must be energy management. Demand switched lighting has long been known as cost saving because using this technique ensures that luminaires are only energized when lighting is needed. Low energy extended period can equally exploit the opportunity of replacing lighting that is wasting energy.

Table 18.1 *Lighting applications*

Application	*Observation, activity and benefits*
General	• Lighting more than 10 years of age. • Fluorescent tubes with bayonet cap connections. • Discoloured plastic diffusers. These should all be replaced. A refurbishment programme should produce 60% energy savings and a payback in less than 18 months. An improvement in the quality of lighting is expected and there is a high probability of major cost savings. • Ordinary coiled coil filament light bulbs. Renew with compact fluorescent lamps in the existing fitting or change the fittings in total. Replace with PAR-E or use low voltage tungsten halogen lighting or metal halide discharge lighting. There should be an immediate cost saving of 75% plus reduced maintenance through longer lamp life. Overall lighting costs can be halved at low capital spend.
Industrial	• Mercury fluorescent lamp bulbs (MBF type) blue mercury lights. • Mercury/tungsten 'blended' lamps (MBTL type). • Aged 8 foot 125 W fluorescent fittings. • High wattage light bulbs (300 W to 1500 W filament lamps). These should all be replaced. For high bay applications they are to be changed to high pressure SON units. For low mounting heights they are to be replaced with either high pressure sodium SON lights or more modern fluorescent fittings. Benefits are in SON providing satisfactory colour rendering in most industrial applications and also being energy effective. High frequency electronic or low-loss fluorescent gives energy costs almost as effective as SON plus excellent colour rendering.
Commercial	• 2 foot 40 W fluorescent fittings. • Fluorescent fittings with opal diffusers. • Warm tone fluorescent lamps. These are best replaced with modern louvred or high performance prismatic lens fittings using power saving triphosphor lamps. High frequency electronic or low cost ballasts are particularly effective. Metal halide and SON uplighters can also be used in some applications. The benefits should include energy savings of 30–45% with an improvement of lighting quality. There will also be less glare, flicker and hum with easy starting and good colour rendering.

PAR 38 sealed beam reflector lamps and other reflector lamps can be replaced with PAR-E or low voltage tungsten halogen lighting or metal halide discharge lighting. This will give energy savings of 30–75% for an equivalent lighting performance.

Outdoor

- Filament lamps.

  These can be replaced with SON high pressure sodium lamps or compact fluorescent luminaires. This would give an energy saving of 75–85% and lower maintenance through longer lamp life and improve the quality of the lighting.

Floodlighting

- High wattage filament lamps and tungsten halogen floodlights.

  These are best replaced with SON high pressure sodium or mercury discharge lighting.
  This will provide energy savings of 60–80% plus superior light forms.

Lighting control

- Large areas that are controlled by badly designed switches controlling excessive luminaires.

  These systems should be replaced by modern light control systems as these can reduce electricity consumption by 25–60% and provide far better working environments.

In practice the easiest way to save energy from a lighting installation is to turn off the lights when no one is present or at those times when sufficient natural light is available. The problem is that personnel cannot be relied on to turn them off.

Advancements in energy management and lighting control systems using digital control techniques can now be installed to provide the required functions such as:

- Provide light only when people are present.
- Provide light only when natural daylight is insufficient.
- Provide only the correct level and quantity of light.
- Provide the ability to reconfigure lighting systems to satisfy changing needs and without any cause to amend expensive equipment.
- Provide artificial light in an efficient way.
- Provide control for the individual.

By using passive infrared presence detection the lighting can be turned on when an individual enters an area and then turn them off when the area is vacated. The lights may also be turned down as the daylight increases maintaining the required light levels and saving energy in areas adjacent to windows.

In order to achieve this goal the PIR detectors may be fitted within each luminaire or mounted in a particular location that enables them to control a bank of luminaires.

A true energy management system that provides automatic control of lighting would be microprocessor based. This would have light sensing photoelectric cells that detect the level of illuminance and by a method of sequence switching would turn off programmed luminaires in response to a preset value. Further technology can lead to the dimming of lamps to go down to a value of some 25% of the maximum level. These control systems that are energy conscious and give a degree of security give feedback signals to a comparator or controller that can work in conjunction with light dependent resistors. In terms of security these provide a measure of reassurance light in preference to no light being available.

Whereas the true energy management system is intended for office complexes there has long been a cost conscious range of products available to carry out a measure of domestic security lighting for indoor premises and to give the appearance that persons are present. These components are themselves energy conscious as the lighting is only available for short time scales in preference to leaving lights activated when premises are vacated. Popular are timer switches that can interface with a lamp via a socket outlet but purpose designed timers can replace manual light switches to randomly select lights to a number of varied programmes.

Photocells and dimmer switches intended for the domestic market are not difficult to install and can be used for internal applications. Dusk–dawn light switches that plug into existing light sockets will switch off automatically as light levels rise. These are found in different guises for incandescent lights, fluorescent or low energy luminaires and can also be random in operation.

Indeed there are countless inexpensive security lighting products that satisfy energy management concerns and provide a low level of security protection. Included are sound activated switches, copy switches that will learn and then copy switching patterns and command centres that enable control of external security systems and operation of lights and auxiliary equipment. In the same way that occupancy detectors and automatic 'night lights' are promoted as security lighting devices that also save energy they can also be used with confidence in low risk applications.

From this we can safely say that security lighting can be a part of energy management and a serious security solution for a low risk residential application or a large industrial site. In the domestic environment many of the products may not be high technology yet they still have a role to play. Nevertheless there must still be attention to detail because if the lighting does not stop intrusion or enables perpetrators to be identified it has failed in its task.

19 Integrated systems and central station monitoring

There was a time when the different disciplines and technologies in security were deliberately kept apart and all systems retained separate identities. Therefore there was no overlap between intruder alarm, access control, CCTV and lighting technologies and specialists would only show an interest in the various subjects as standalone techniques. However, as the security industry has progressed and grown so has the need for integration of systems to the extent that it is now big business. In addition we now have intelligent management systems that bring together all of the technologies. These may often be monitored from a central station or alarm receiving centre (ARC) that provides 24 hour cover. This is necessary if complex security solutions are to be addressed. Manufacturers are designing products and systems for the future and putting in place infrastructures to meet the needs of innovative networks. Software is being created for integration because the future of the security industry will be different to today's market place as standalone technologies will become increasingly replaced by systems involving lighting, security, safety, site management, personnel management, environmental control and home automation.

We tend to see integration and home automation in the same general sense because they all involve the control of a number of systems within one framework that may be performed from a remote location. The interests will lie in the selection of security functions together with the control of light, heat, environmental conditions and the links that they may have with appliances or equipment contained in any building.

The simplest form of integration is to have completely separate systems initially but then to link these by a simple interface such as a relay and a pair of wires. Examples of this are lighting and indicator networks that are energized by the signalling from an access control or intruder alarm system that secures an area. Figure 19.1 shows as an example, a lighting and indicator network that is energized by the output from an access control system that has had a secured protected door violated. In this case an intruder alarm is further connected and using a zone programmed as auxiliary which is a circuit that is monitored at material times and when opened will activate outputs that are programmed for special duties. In this figure when a door controlled by the access system is forced, the signalling from the separate intruder system is used to supplement the demand lighting. A neon, an LED and a buzzer used alongside each other provide indication of the alarm condition with the time of the activation being determined by the programmed timeout of the access alarm condition.

It can be noted that a number of different relays are used to connect the various systems to achieve an integrated network. A double pole relay is used to connect a CCTV system in the network and to select a specific camera by connecting the appropriate input on the switcher, quad or multiplexer to bring up the video display. The lighting is controlled by a different and particular relay with the switching contacts rated for mains duty.

From this we can say that the simplest form of integration and automation is achieved by separate systems that are only linked by normally open or normally closed switching. This allows one system to control or to use the facilities of another by a pair of wires. By this technique the more specific and complex systems, although integrated, still retain their unique identity which ensures that where such systems as fire alarms are required by law their control and indicating equipment remains exclusive.

Progression in the world of integration and home automation has led to the introduction of bespoke packages that generally comprise an end station which can have its detection circuits wired in a number of ways. These are programmed with the attributes required so that the circuits are able to function in a variety of security modes. To the

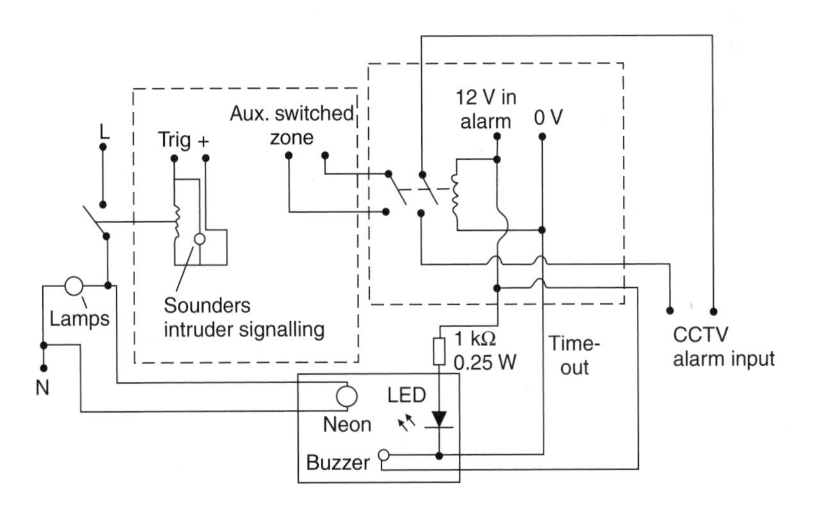

Figure 19.1 *Simple two wire integration*

end station are connected the remote keypads or door readers and the outputs are selected depending on the signalling states. These packages are known as integrated security control and management systems and can be used for a small private house yet be expanded to cover a large industrial or commercial site. The zones are essentially intruder, personal alarm and fire plus an ability to read tags with additional applications of call with mimic display, CCTV switching, lighting control, hold-up alarm and deterrent warning. The future holds good potential for such packages that are available as a kit so the wiring between the components is clearly defined. Therefore they are simple to install, use, expand and service. The controller may also have built-in diagnostic facilities with volt/current/resistance meters for the electrical service. The essential options of adjustable timers and full reporting can be accommodated with multiple zone groups controlled by readers. Heavy-duty relays allow further appliances and equipment to be automated.

Progression in security will outdate standalone systems and we shall see advances in the employment of these management systems. Equipment is becoming more available for all of the duties at the different levels and included is the capacity for communications and remote signalling with service history and reporting.

Certainly the joining of systems does allow the sharing of information. This can range from simple control by means of a pair of wires as shown in Figure 19.1 to the more complex end of the scale such as the formation of a single administration point for all of the systems in a building to include the security lighting. In the latter case software is created so that the user interface reflects all of the systems to include the building control applications alongside the security functions. In this way there are intelligent gateways from product to product although it is best if the systems can still be managed separately with individual maintenance from outside of the core system in the event of failures.

The most practised technique of integrating systems and introducing automation and building control is hardwiring with control wires to a consumer electronics bus local area network. Bus is the name given to a set of lines in a microprocessor circuit. The buses of a microprocessor system consist of lines that are connected to each other and every part of the system so that signals are made available at many chips simultaneously. They are widely used in digital electronics for devices to carry out actions described as controlling, listening and talking. With the bus system a number of devices can be connected to the medium and in general any device can transmit and receive messages to or from one another. Once the central control unit has been installed the output control codes are selected and downloaded to a computer program. The system itself uses standard security detectors so their methodology and resistance to false alarms is guaranteed. The nodes have unique addresses so substitution is not possible and therefore the whole security level remains high. To improve reliability and fault analysis even further subnetworks can be configured using hubs. The main control board will be found to be equipped with the telephone interfaces, inputs and relay outputs for the connection of the different forms of equipment and appliances. This is shown in Figure 19.2.

As the security industry has expanded to embrace integrated products and the control of building management systems there has been an increasing role to play for central stations. Although these alarm receiving centres are becoming increasingly prominent they would not normally be used to monitor security lighting systems in their own right but to accept transmissions from networks that include such systems as part of an integrated role. With enhanced communications and the expansion of digital networks the verification that security lighting has been selected can be confirmed for high security risk applications. This can readily be done over the remote signalling network in conjunction with other input/outputs of other security and building management services. For true verification security lighting is normally supported by CCTV images transmitted over the telephone network link to a permanently manned station.

It becomes apparent that although security lighting can form part of a much greater industry it is without doubt even in its own right a tremendous deterrent to crime. In addition we can observe that although simple security measures reduce the incidence of crime as we introduce more sophisticated technologies and systems it is reduced to an even greater degree. Security lighting can of course not only be customized to suit any application and reduce applied crime and intrusion but it may also lead to overall cost savings in energy when considered as a management technique.

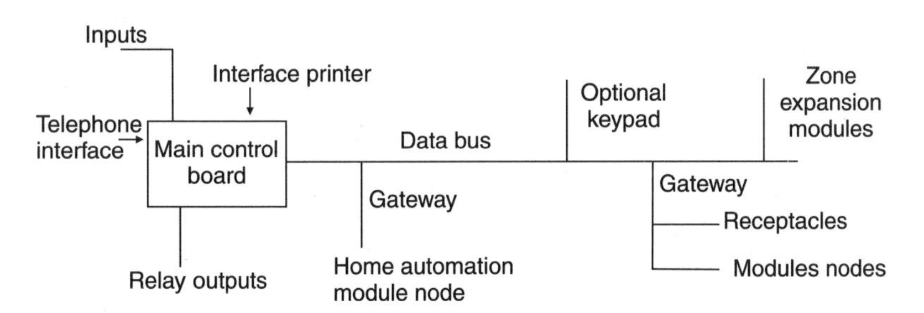

Figure 19.2 *Security and home automation. Data bus configuration*

A further aspect of note is that in relation to psychological protection because this forms the basis of deterrence. It is not normal that an intruder would consider attacking well-protected and illuminated areas if more vulnerable targets existed. Lighting is ideal because it shows that security is being provided especially when it forms perimeter protection. It keeps perpetrators at a psychological disadvantage because of its very strategic overt role and ensures that intruders remain uncertain as to whether they may have been observed. These intruders or criminals also feel insecure believing that they could be surprised by other protective means, security devices or indeed by persons as a result of their detection.

Unfortunately there are suggestions that the levels of criminal activity, particularly those associated with mugging and robbery, are liable to increase. However, we can safely say that good quality security lighting has without any doubt got a major role to play in countering these across the spectrum of security concepts.

20 Reference information

20.1 Environmental protection

All components for both indoor and outside use are governed by their ability to withstand extremes of climate, changing ambient temperature limits, the weather and the ingress of liquids and dusts. The degree of protection is indicated by the letters IP followed by two characteristic numerals. Reference can be made to standards BS 5420, IEC 144 and IEC 529.

The first numeral indicates the protection afforded against the ingress of solid foreign bodies and the second the protection against the ingress of liquids (see Figure 13.1).

Table 20.1 *The NEMA system*

North America and those countries that are influenced by American standards use the NEMA system with the classifications contained in publication No. ICS-6. In general those that we are interested in are:

Type 1. General purpose, indoor.
Intended to prevent accidental contact of personnel in areas where unusual service conditions exist. Protection is afforded against falling dirt. The enclosures may or may not be ventilated.

Type 2. Drip-proof, indoor.
Intended to protect the enclosed equipment against falling non-corrosive liquids and falling dirt. These enclosures may have provision for drainage.

Type 4. Watertight and dust-tight, indoor and outdoor.
Intended to protect against splashing or hose directed water, seepage of water or severe external condensation. This level of protection corresponds generally with IP54.

Type 6. Submersible, watertight, dust-tight and sleet resistant, indoor and outdoor.
Intended where occasional submersion is encountered. This classification is generally as IP67.

Type 13. Oil-tight and dust-tight, indoor.
Intended to afford protection against lint and dust, seepage, external condensation and the spraying of oil, water and coolant. This classification is generally as IP65.

It should be appreciated that many goods are supplied in housings and are coded to indicate the classification when gaskets are correctly seated and conduits etc. sealed.

The letter W may sometimes be used as a supplement. It is added to show that particular features exist under specific weather conditions and are endorsed in manufacturers' data.

20.2 Multiplication factors

Factor	*Prefix*
1 000 000 000	Giga (G)
1 000 000	Mega (M)
1000	Kilo (k)
100	Hecto (h)
10	Deca (da)
0.1	Deci (d)
0.01	Centi (c)
0.01	Milli (m)
0.000 001	Micro (μ)
0.000 000 001	Nano (n)
0.000 000 000 001	Pico (p)

20.3 Common multiples

Unit	*Multiple*	*Value*
Ampere	Milliampere (mA)	1/1000 ampere
Ampere	Microampere (μA)	1/1 000 000 ampere
Volt	Millivolt (mV)	1/1000 volt
Volt	Microvolt (μV)	1/1 000 000 volt
Ohm	Kilohm (kΩ)	1000 Ω
Ohm	Megohm (MΩ)	1 000 000 Ω

20.4 Standards, regulations and codes of practice

Overview

In all industries there is some form of legislation that must be adhered to in order that safe working environments are produced. In some countries national legislation will be produced but in others the laws that interpret safety issues will vary. Lighting is controlled in much the same way.

Security lighting does not have any specific standards that govern its content because of the diverse and wide nature of its activities when it is installed as a system. However, the majority of individual components that it employs do have their own particular standards and the general requirements for electrical installations apply to the installed security lighting system in the same way that they relate to any form of electrical network.

For these reasons we are only able to include reference information as related to security lighting as an overall concept. However, the listings do enable the reader to understand the very nature of the requirements that apply and how they must be satisfied.

The information that follows is therefore presented in the form of references and suggested further reading.

BS/EN – European.
British Standards – BS.
EN – European Standards.
BS EN ISO – Harmonized.
IEC – International Electrotechnical Commission.

BS EN 50133: Alarm systems. Access control systems for use in security applications.
BS EN 50133-1: System requirements.
BS EN 60598: Luminaires.
EN 50131.1: Alarm systems. Intrusion. System requirements.
EN 50131.6: Alarm systems. Intrusion. Power supplies.
EN 50132.1: CCTV. System requirements
EN 50132.7: CCTV. Application guidelines.
EN 50136.1.3: Alarm transmission systems for digital PSTN.
EN 55 015: Radio interference limits (fluorescent lamps).
BS 3116: Part 4: Specification for automatic fire alarm systems in buildings. Control and indication equipment.
BS 4533: Luminaires.
BS 4737: Part 1: Intruder alarm systems in buildings. Specification for installed systems with local audible and/or remote signalling.
BS 5345: Code of practice for selection, installation and maintenance for electrical apparatus for use in potentially explosive atmospheres (other than mining applications or explosive processing and manufacture.
BS 5420: Degrees of protection of enclosures for low voltage switchgear and control gear.

BS 5501: Electrical apparatus for potentially explosive atmospheres.
BS 5489: Road lighting.
BS 5839: Fire detection and alarm systems in buildings.
BS 5979: Code of practice for remote centres for alarm systems.
BS 6301: Requirements for connection to telecommunication networks.
BS 6360: Conductors in insulated conductors and cables.
BS 6467: Part 2: Electrical apparatus with protection by enclosure for use in the presence of combustible dusts.
BS 6800: Specification for home and personal security devices.
BS 7042: Specification for high security intruder alarm systems in buildings.
BS 7671: Requirements for electrical installations. Issued by the Institute of Electrical Engineers as the IEE Wiring Regulations.
BS 7807: Integration of fire and security systems.
BS 8220: Guide for security of buildings against crime. Part 1: Dwellings. Part 2: Offices and shops. Part 3: Warehouses.
BS EN ISO 9000: Quality management in quality assurance standards.
IEC Document No. 123-93: International lamp coding system.

Regulations and codes of practice

Building Regulations. Approved Document B.
Code for interior lighting. ISBN 0 900953 64 0.
Health and Safety at Work Act 1974.
Electricity at Work Regulations 1989.
The Factories Act 1971.
The Memorandum of Guidance on the Electricity at Work Regulations 1989.
Electromagnetic Compatibility Regulations 1992. Statutory Instrument SI 2372.
CIBSE LG1: Lighting Guide. The Industrial Environment.
CIBSSE: Lighting for Offices.
Health and Safety Executive HSE 253/3: Disposal of discharge lamps.
BSIA. British Security Industry Association 107: Access control systems. Planning, installation and maintenance.
BSIA. British Security Industry Association 109: CCTV systems. Planning, installation and maintenance.
BSIA. British Security Industry Association 195: EMC guidelines for installers of security systems.
NACOSS. NACP 11: Supplementary code for the planning, installation and maintenance of intruder alarms.
NACOSS. NACP 13: Code for intruder alarms for high security premises.
NACOSS. NACP30: Code for planning, installation and maintenance of access control systems.

Index

VCTC Library
Hartness Library
Vermont Technical College
Randolph Center, VT 05061
DISCARD